UNIVERSITY OF WINCHESTER
LIBRARY

Martial Rose Library
Tel: 01962 827306

3 0 SEP 2013

To be returned on or before the day marked above, subject to recall.

KA 0378999 3

SUNY series on Religion and the Environment
Harold Coward, editor

PLANTS AS PERSONS

A Philosophical Botany

MATTHEW HALL

UNIVERSITY OF WINCHESTER
LIBRARY

Cover art, "Betula pedant," courtesy of Mairi Gillies www.mairigillies.com 2010.

Published by State University of New York Press, Albany

© 2011 State University of New York

All rights reserved

Printed in the United States of America

No part of this book may be used or reproduced in any manner whatsoever without written permission. No part of this book may be stored in a retrieval system or transmitted in any form or by any means including electronic, electrostatic, magnetic tape, mechanical, photocopying, recording, or otherwise without the prior permission in writing of the publisher.

For information, contact State University of New York Press, Albany, NY
www.sunypress.edu

Production by Kelli W. LeRoux
Marketing by Anne M. Valentine

Library of Congress Cataloging-in-Publication Data

Hall, Matthew, 1980-
Plants as persons : a philosophical botany / Matthew Hall.
  p. cm. — (SUNY series on religion and the environment)
Includes bibliographical references (p.   ) and index.
ISBN 978-1-4384-3429-2 (hardcover : alk. paper)
ISBN 978-1-4384-3428-5 (pbk. : alk. paper) 1. Botany—Philosophy. I. Title.

QK46.H35 2011
580.1--dc22

2010018516

10 9 8 7 6 5 4 3 2 1

For Val

Contents

	Acknowledgments	ix
	A Philosophical Botany	1
1	THE ROOTS OF DISREGARD: EXCLUSION AND INCLUSION IN CLASSICAL GREEK PHILOSOPHY	17
2	DOGMA AND DOMINATION: KEEPING PLANTS AT A DISTANCE	37
3	PASSIVE PLANTS IN CHRISTIAN TRADITIONS	55
4	DEALING WITH SENTIENCE AND VIOLENCE IN HINDU, JAIN, AND BUDDHIST TEXTS	73
5	INDIGENOUS ANIMISMS, PLANT PERSONS, AND RESPECTFUL ACTION	99
6	PAGANS, PLANTS, AND PERSONHOOD	119
7	BRIDGING THE GULF: MOVING, SENSING, INTELLIGENT, PLANTS	137
	Recreating a Place for Flourishing	157
	Notes	171
	Bibliography	209
	Index	229

Acknowledgments

A number of people have assisted with the research for this book, and it is a pleasure to thank them here.

Deborah Rose, Val Plumwood and Libby Robin introduced me to the *ecological humanities* (www.ecologicalhumanities.org) and helped me to develop the ideas and themes that run throughout the book. Graham Harvey guided me through the field of religious studies and Will Tuladhar-Douglas, Cadaran, Phillip Carr-Gomm, Damh, Sally Alsford, Potia, Miranda Morris, and Siusaidh Ceanadach provided important input into the chapters which deal with plants in religious traditions. Tony Trewavas and Thom Van Dooren kindly discussed the concepts behind their own work. Fritz Detwiler, John Grim, and Freya Mathews and two anonymous reviewers provided valuable comments and corrections to early versions of the manuscript.

Thanks also go to the SUNY Press team: Nancy Ellegate, Kelli Williams-LeRoux, Thomas Goldberg, and Anne Valentine.

The biggest thanks remain for the love and support of my family and friends.

A Philosophical Botany

> How should we speak to trees, how should we treat the trees, other animals and each other that all of us can live and live at peace?[1]
>
> —Erazim Kohák

Replanting Nature

Most people are aware that human beings are harming nature. Every iconic picture of a dying rainforest, a slaughtered tiger, or a poisoned river rams home the fact that human relationships with the natural world are increasingly destructive. In some of the strongest analyses of our environmental crisis, it has been instead that human hyperseparation from the natural world has entangled us in what conservation biologists recognize as the sixth great extinction crisis—a crisis of death that is human made. Environmental philosopher Val Plumwood has put forward the idea that the prevailing Western culture has created a human-nature dualism.[2] In this worldview, nature is constructed as radically different from the human, and human culture is radically separated from it. Plumwood argues that Western worldviews in particular render nature as an insignificant Other, a homogenized, voiceless, blank state of existence, a perception of nature that helps justify domination of the Earth.[3]

Largely because it is depicted as devoid of the attributes which require human attention—such as mentality, agency, and volition—nature is left out of the sphere of human moral consideration. In the words of the recent UN GEO4 report, the resulting behavior toward the natural world constitutes an assault on the global environment that risks undermining biospheric integrity.[4] An appropriate response to the swathe of environmental problems created by human beings must be to develop less destructive, more respectful, harmonious relationships between humans and nature. Yet, the concept of nature is somewhat elusive and homogenous.

To the postmodern deconstructionist, nature is a provocative term, a human construct, created only in its situated opposition to the human realm of culture.[5] In the physical sciences, nature is thought of in terms of universals and inviolable laws. The physicist and the astronomer form their idea of nature from celestial bodies, and their governing forces. Back on Earth, the cultural geographer David Harvey perceives nature in terms of dialectics, as a series of processes and flows.[6] This idea of nature as a system of transfers has much in common with contemporary ideas in the ecological sciences. Within the realm of environmental philosophy, process-based understandings of nature are often advocated. Freya Mathews regards nature as the absence of abstract design, as "whatever happens when we, or other agents under the direction of abstract thought, let things be."[7]

Despite the abundance of philosophical and everyday references to nature, it is clear that within environmental studies, "nature has remained a largely undifferentiated concept, its constituent parts rarely theorized separately."[8] Therefore, a logical response to the challenge of renewing ethical relationships with nature in a time when much of the nonhuman world is threatened by human activity is to theorize human-nature relationships in terms of heterogeneity. We must take Plumwood's two major tasks for humanity, "(re)situating humans in ecological terms and non-humans in ethical terms" and apply them in terms of a separately theorized nature of diversity, abundance, and individual (as well as collective) presence.

Using insights from biology, such activity has been proceeding for some time in the broadly defined discipline of animal studies. For decades, animal rights theory has been directly concerned with establishing more ethical relationships with animals. Leading animal rights theorists such as Peter Singer, Tom Regan, and Gary Francione have used an understanding of the sentience and subjectivity of animals to argue for their moral consideration.[9] Ethologist Mark Bekoff has also pioneered this approach in zoology. A leading voice for the ethical treatment of animals, Bekoff directs his research to maximize human recognition of animals as fascinating, complex, social beings—autonomous individuals that fully deserve human moral consideration.[10] Such detailed biological knowledge of animal physiology and behavior has prompted a number of wider animal-human studies that have aimed at reestablishing human-animal relationships on more moral terms.[11] In view of the major tasks for humanity, these studies have begun to question the dualism of humans and nonhumans and have begun to open up the possibility of moral consideration for nonhumans.

While such studies of moral consideration for animals proliferate, studies that focus on arguing for the moral consideration of plants are rare. Yet, recognizing the need for more ethical human-nature relationships and the need to theorize the constituents of nature separately, we must also acknowledge that the largest component of a nature composed of nonhuman beings is not composed of animal life. In the Earth's deserts and on her mountainous peaks, much of the nonhuman world is composed of rock. In her seas, lakes, and rivers, the biggest nonhuman presence is water. However, in the majority of places that are inhabited by people—even within towns and cities, particularly in Europe and North America—plants dominate the natural world.

Most places on Earth which contain life are visibly *plantscapes*. Whether they walk in human transformed habitats or in wilderness, human beings are far more likely to encounter plants than any other type of living being.[12] In fact, the bulk of the visible biomass on this planet is comprised of plants.[13] It is a fact that in most habitable places on Earth, being in the natural world first and foremost involves being amongst plants, not amongst animals, fungi, or bacteria. Although fungi, bacteria, and animals are important for sustaining natural processes, plants are the most abundant form of life in nature that humans encounter.[14] Importantly, both directly and indirectly, it is the visible presence of this plant biomass which enables the presence and continued existence of human beings.

Within the imprecision of the term *nature*, the global dominance of the plant kingdom is seldom recognized. In a plant-dominated biosphere, it is possible that nature has become so amorphous and peripheral because of the way that plants (synonymous with nature) are themselves perceived. A long overdue study on human-plant perceptions and relationships is crucial therefore for understanding how we treat the natural world.[15]

Here I will base such a study on an extended investigation of the cultural and philosophical orientations that are critically important in human considerations of the natural world. Philosopher Erazim Kohák has coined the term *philosophical ecology* to articulate the need to incorporate these considerations within environmental discourse.[16] This text applies the idea of a philosophical ecology to the botanical world and avers that a study of different cultural-philosophical perceptions of the plant kingdom is crucial for developing more ethical relationships with the plant kingdom. By examining a variety of contrasting cosmologies, philosophies, and metaphysical systems that deal explicitly with plants, one of the main aims of this study is to uncover how and where plants are placed within a variety of human worldviews. In doing so, it will dissect how these plant philosophies determine the overriding relationships that human beings have with the plants they live amongst. An important aspect of this task is an extended analysis of the processes by which plants find themselves included or excluded within the realm of human general and moral consideration.

The task is to survey a number of plant knowledges in order to uncover the most appropriate human rendering of plant life. At a time when many plant species and indeed the natural world itself, are threatened by human activity, this study also aims to locate the most appropriate human behavior toward plants. This dual approach is again set within the parameters of Kohák's philosophical ecology. Its thrust follows Kohák in his search for the most appropriate "manners of speaking" rather than looking for a "positive description of a univocal 'metaphysical' reality."[17]

Within the context of an anthropogenic ecological crisis, the choice between different modes of perception and action is an important one. Human life is contingent upon the existence of plants. Throughout this work, I repeat the assertion that our general, Western, view of plants as passive resources certainly plays a significant role in our ecological plight. Finding a more appropriate way of approaching plant life could underpin a mode of human action that maintains, rather than threatens, biospheric integrity.[18] By surveying a number of cultural and philosophical sources, the aim of this work is to incorporate cultural and metaphysical influences into botanical studies—perhaps the rendering of a more *philosophical botany*.

Throughout the pages of this book, I search for the most appropriate behavior toward plants in a time of impending ecological collapse. This broad approach resonates with Freya Mathews's idea, that we "must draw on as wide a range of cultures as possible" in order to develop complex ethical solutions to environmental problems.[19] The point is not to appropriate other knowledge systems, nor elevate one worldview as *best*, but to investigate a number of worldviews in order to generate ideas and strategies for more appropriate ecological behavior in a Western context. As Eliot Deutsch makes clear, we don't turn to different cultural ideas "for a better scientific understanding of nature . . . but for different ontological perspectives and moral ideals that might influence our own thinking."[20]

By working with a number of small case studies, this survey will construct a meta-narrative, examining the influential factors and the processes that determine how plants come to be placed within particular worldviews. When questioning human perceptions of and behavior toward the plant world, it is clear from the outset that Erazim Kohák is onto something. Cultural-philosophical ideas strongly influence human interactions with the plant kingdom, and humanity possesses a multitude of different ways of thinking about and acting toward plants. The following chapters analyze contrasting modes of perceiving and behaving toward plants. For clarity, I have split these modes of perception into broadly defined *philosophies of exclusion* and *philosophies of inclusion*.

Perspectives and Processes of Exclusion

The first three chapters tackle the marginalization of plants using the themes of radical separation, zoocentrism (an animal-centered outlook), exclusion, and hierarchical value ordering. These chapters argue that these notions predominate in Western discussions of plant ontology. In the terms of the scholar Mikhail Bakhtin, such approaches to life can be broadly classified as monological. For Bakhtin monologue does not recognize the voice or presence of the other; it "is finalized and deaf to the other's response, does not expect it and does not acknowledge it any decisive force. Monologue manages without the other. . . ."[21]

The question of the human marginalization of plants has started to receive some attention in the botanical sciences. In a seminal study of plant physiology, vegetation ecologist Francis Hallé starts with the presumption that human beings are ignorant of the biosphere's plant life. He contends that the majority of people are "generally poorly acquainted with plants, looking down on them or simply ignoring them."[22] In the same vein, botanists and environmental educators Wandersee and Schussler have written of the phenomena of *plant blindness*, the literal ignorance of plants by human beings and their spontaneous preference for animal life.[23] According to the authors, some of the symptoms of this widespread "disease" are:

- Failing to see, take notice of, or focus attention on the plants in one's life.
- Thinking that plants are simply the background for animal life.
- Overlooking the importance of plants to human life.
- Misunderstanding the differing time scales of plant and animal activity.[24]

Hallé postulates that human beings place little value on plant life because of a prevailing zoocentrism or anthropocentrism. This stems from the fact that human beings do not as readily identify with plants as with animals and that humans lack a spontaneous appreciation of the plant's physiological workings.[25] Although we are surrounded by plants and cultivate them fanatically in our gardens, he regards plants as being in a state of "absolute otherness" to human beings, by which he means that plants operate their lives in vastly different ways to the animal *Homo sapiens*. Hallé has attempted to correct this situation by increasing the knowledge of plant physiologies and life histories, focussing on the ways in which plants differ significantly from the human and the animal.

From a similar understanding of the problem, Wandersee and Schussler have attempted to explain the ignorance toward plants and a prevailing zoocentrism through an analysis of the psycho-optical prejudices in animals.[26] Lack of knowledge of plants, the general similarity of plant surfaces and textures, the lack of movement in plants, and the fact that plants do not prey on humans are all put forward as possible reasons for the phenomenon of *plant blindness*.[27] In a more recent paper, Wandersee and Clary make it clear that they regard the neglect and ignorance of plants to be symptoms of an underlying physiological bias:

> In challenging the conventional wisdom, we have proposed that those first three behaviors zoocentrism, zoochauvinism, plant neglect are actually *symptoms* of the default *condition* of plant blindness arising from how the human eye-brain system typically processes and attends to visual information. . . .[28]

While these botanists have strongly articulated the problem of plant ignorance, their analyses of its causes remain incomplete. They have identified the problem of a zoocentric attitude toward plants, yet the reasons they provide for its existence are potentially misleading. By positing a physiological basis to this problem, they implicitly suggest that such a zoocentric attitude is in a sense *natural* and *inevitable* for all human beings. I argue that the marginalization that characterizes Western thought is neither natural nor inevitable. Zoocentrism does not emerge from physiology, but is largely a cultural-philosophical attitude. The fundamental mistake here is the assumption that this zoocentrism found in Western society pervades all cultural ideas and actions toward the plant kingdom. This closed stance leaves little room for the recognition of alternative approaches.

The opening three chapters deal with broadly Western streams of thought, and each chapter clearly demonstrates the predominance of zoocentric perspectives. Here, my key arguments are that the insignificance of plants in contemporary Western society identified by Hallé et al. is partly generated from a drive toward separation, exclusion, and hierarchy. My analysis focuses on the bases for such ideas, the processes by which they have been solidified, and the outcome for human behavior toward plants.

The material in the first three chapters agrees with Hallé's recognition that Western societies have a predominantly zoocentric vision, but differs in its claims that zoocentrism is a deliberate philosophical strategy for marginalizing and excluding plants. Zoocentrism is a *method* for achieving the exclusion of plants from relationships of moral consideration. For want of a better term, it is a *political* tool in an exclusionary process in which "the Other becomes a negative necessity, that which must be set apart and kept apart for one's own self of collective self to be sustained."[29] Zoocentrism thus helps to maintain human notions of superiority over the plant kingdom in order that plants may be dominated. It is a crucial dualising force, responsible for depicting plants as inferior beings and as the natural base of a human-dominated hierarchy.

Along with the dualisms identified by Plumwood,[30] constructing a rigid hierarchy in which those at the top have more value is fundamental to encouraging radical separation of different groups and to justifying a logic of domination by the upper echelons of the hierarchy.[31] In the field of social ecology, Murray Bookchin has identified the construction of human hierarchies as the justificatory basis of dominance by one human group over another.[32] In ecofeminist theory, Karen Warren has identified value-hierarchical thinking as part of an oppressive conceptual framework that "functions to explain, maintain, and 'justify' relationships of unjustified domination and subordination."[33] This separation and value-ordering is a crucial part of the general drive toward excluding plants from human consideration. This trend is very important to uncover and ultimately redress, for exclusion is "an act of intellectual violence; and it is the attitude that drives collective and systematic physical violence."[34] The intellectual violence of backgrounding plants and denying their sentience can be said to underpin the "occupation, appropriation, and commodification" of the plant kingdom and thus the wider natural world.[35]

In considering Western attitudes toward plants, this hierarchical ordering based upon the construction of exclusionary, "oppositional value dualisms" is predominant in some of the Western world's most influential, penetrating philosophies.[36] Chapter 1 deals with the construction of hierarchies in the natural world, and the dualistic treatment of plant abilities and faculties within the philosophical tradition of the ancient Greeks. The analysis of these constructions begins with Plato, who defined plants from a dualistic zoocentric perspective and asserted that plants were created expressly for the use of human beings. This approach was perpetuated by Aristotle. Aristotle judged the abilities of plants on the basis of what he had observed in animals, rather than considering plants on their own terms. Aristotle constructed a hierarchy of life with plants placed firmly at the bottom. Underpinning this hierarchy, plants were rendered *radically* different from animals, regarded to lack the faculties of sensation and of intellect. Such hierarchical ordering demonstrates a drive toward separation; one that is based upon removing continuities from plant and human life. This is a stance which solidifies exclusion.

One of the key features of Chapter 1 is that it explores the effect of perspective and intent on the human approach to plant life. It contrasts Aristotle's hierarchical ordering, his drive toward separation and exclusion by removing human-plant continuities, with the work of his pupil Theophrastus. Examining the Theophrastean perspective, Chapter 1 reveals that this stance of exclusion is neither natural nor inevitable. It is human intent, rather than the differing physiology of plants which creates radical exclusion.

In contrast to Aristotle, the work of Theophrastus attempts to treat plants on their own terms and emphasizes their relatedness and connectedness to humanity. Such an approach to plant life is very similar to that found in ancient Greek mythology and the surviving fragments of pagan traditions across Europe.[37] It is apparent in the work of Theophrastus that rather than exclusion, his orientation was toward inclusiveness and consideration. The result of this difference in intent is phenomenal. Instead of regarding plants as passive beings lacking sensation and intellect, Theophrastus related to plants as volitional, minded, intentional creatures that clearly demonstrate their own autonomy and purpose in life. For Theophrastus, plants demonstrated their own purpose and desire to flourish through their choice of habitats and the production of seed and fruit.

Chapter 2 explores some of the reasons why the predominant Western treatment of plants more closely resembles that of Aristotle rather than Theophrastus. Although he had a great impact on the development of large parts of botany, it is unfortunate that Theophrastus's philosophical orientation was not followed or developed. This chapter deals with disappearance of Theophrastus's insights, the predominance of Aristotle's hierarchical philosophy and the analysis of zoocentrism in botanical history in greater depth. In particular it examines how readily a hierarchical approach to plants has been retained in the botanical sciences, with plants increasingly excluded on the basis of ancient zoocentric philosophy. One of its most important points is that the systematic devaluation of such a large part of the natural world had been occurring long before Cartesian philosophy and the rise of an industrial mechanistic atomism.[38]

Chapter 3 continues the theme of hierarchies and looks at the interpretation of plants within Christian theological sources, specifically biblical material and the writings of prominent theologians. It is clear that biblical texts also construct plant life as radically different to humans and animals. In the biblical creation stories, there is a further drive toward emphasizing the differences and rejecting the continuities between plants and humans. Although plants display the characteristics of other living beings such as growth and death, they are not considered to be alive. While the possibility exists for a more inclusive approach to plant life on a number of criteria, they are instead separated from the rest of the living world on the basis that they lack the "breath of God." This treatment strips the plant world of both life and any possibility of autonomy. As a result, within Christian theological material, relationships with plants can be characterized predominantly as *instrumental* relationships, based upon the usefulness of plants to human beings. Plants are placed at the bottom of a hierarchy of the natural world and are excluded from human moral consideration.

In the writings of later theologians, the vitality of plants is accepted but the hierarchical view of life is continued, maintaining the instrumental mode of human-plant relationships. There is a tension here between the recognition of plants as living beings and the need to kill plants on a daily basis to survive. Rather than acknowledge this killing, and face possible limits to human action, these hierarchies suppress it. They do this by finding other ways to construct *radical* difference in order to render plants as peripherally insignificant, thus furthering the logic of domination. The hierarchy that is presented in biblical creation stories is solidified using similar ideas from the Greek philosophical tradition. In particular, Aristotle's rendering of plants without intellect, was used by Christian theologians to deny plants the possession of a *soul*.

Chapters 1, 2, and 3 demonstrate that the predominant Western relationships with plants are instrumental and hierarchical, and that the drive toward separation is based upon the systematic devaluation of the *lowliest* parts of the hierarchy. Fundamentally, these are the processes that deny moral consideration to plants. Exclusion is both based upon, and furthers, the denial of plant presence and sentience. Ultimately it denies life and death. This is a denial that renders plants as passive entities and which compellingly reinforces separation and difference. In biblical thought, as well as in Plato and Aristotle, hierarchies are built around the issue of use and violence.[39]

Perspectives and Processes of Inclusion

The treatment of, and response to, plant life and death pervades the majority of the following chapters. Chapter 4 links the case studies already outlined with those that deal with inclusion and connection. As well as inclusion and connection, Chapter 4 also introduces the general themes of *heterarchy* and *dialogue*.[40] Like monologue, here dialogue is defined in Bakhtinian terms—principally the recognition of the other's "voice," standpoint, and presence during interaction.[41]

Along with a drive to treat plants on their own terms, these themes of inclusion pervade the remainder of the chapters in this work. Chapter 4 is the longest as it acts both as a counterpoint and a companion to the first three chapters. Containing a number of conflicting viewpoints, bifurcations, and ambiguities—it is also a companion to the chapters which follow. Valuably, it allows examinations of contrasting processes, which lead to diverging attitudes toward the plant world.

In Chapter 4, I turn toward a consideration of Hindu scriptural sources. Although far from exhaustive, even my limited reading of these scriptures demonstrates that plants are not universally subject to hierarchical separation. In important Hindu texts, plants are described as fully sentient beings with their own attributes of mentality. Significantly, in death, the portrayal of reincarnating souls in the *Upaniṣads* ontologically connects the plant, human, and animal worlds. The interpenetration of these existences engenders the recognition that it is possible for human beings to act violently toward all these types of beings. In the case of plants, this manifests in the human ethical ideal of acting nonviolently toward them.

From this broad philosophical basis, a primary bifurcation between Jainism and Buddhism can be detected. Jain philosophy echoes the general approach of the Hindu scriptures and is a practical example of the systematic application of the philosophy of nonviolence in all dealings with the plant world. Jain philosophy is particularly significant for its prominent inclusion of nonhuman interests within the sphere of human consideration. Jainism seeks affinity with plants, thus fostering nonviolence. Significantly it allows plants space to flourish.

In contrast, although Buddhist cosmology is not inherently hierarchical, in some Buddhist schools, a hierarchy has developed that privileges animals over plants. Certain schools of Buddhism have veered way from the recognition of plants as living, sentient beings, and have neglected them in questions of moral consideration. In this analysis, the work of Buddhist scholar Lambert Schmithausen is particularly important for pinpointing the source of this omission. For Schmithausen, plants have been backgrounded in Buddhist philosophy primarily because of the wish to avoid the explicit recognition of violence. This repressed recognition of violence done toward plants is a crucial point. Importantly, Chapter 4 introduces the idea that this process of philosophical devaluation is not confined to the West. From a position of ambiguity on plant violence, a number of Buddhist schools have developed zoocentric criteria for ethical inclusion and have placed plants outside the realm of sentient life.

Interestingly however, the Buddhist tradition also contains a very different philosophical approach toward plants, suggesting that from a plant point of view there is no single Buddhist tradition. Rather than positioning plants as inferior to animal life, scholars within East Asian Buddhism have sometimes come to regard plant life as superior both in capability and worth. This is an important position because it allows a discussion of the subtle turning points that have produced radically contrasting perceptions of plants within the same general metaphysical framework. Again, it is important to question intent. East Asian Buddhist texts demonstrate a more empathetic rendering of plant life because they directly attempt to expand upon the clear evidence for plant sentience. This is an example of an explicit turning away from the established dogma of inferiority. It is open to interpretation however, whether this direct attempt to turn toward other beings is also an attempt to relate with them using appropriate criteria and include them within the realm of human moral consideration.

While a more empathetic approach to plant life appears in East Asian Buddhism, studies on Indigenous knowledges demonstrate that perhaps there are more appropriateways of *relating* with plants.[42] Indeed it is my contention that as they are often directed at living life in appropriate relationships, Indigenous sources provide the most significant material to contrast with worldviews that seek to exclude plants. Drawing on the work of animist scholars Irving Hallowell, Nurit Bird-David, and Graham Harvey, Chapter 4 draws another important contrast. This is between Western backgrounding of plants and the Indigenous peoples who relate to plants as *other-than-human persons*.[43]

Chapter 5 introduces the themes of personhood, flourishing, and kinship. From a basis that all beings are related, many Indigenous peoples regard plants as beings that possess awareness, intelligence, volition, and communication. Plants are regarded as beings that are capable of *flourishing* and of *being harmed*.[44]

Plants are of course acknowledged as being different from human beings. They have different ways of going about their lives and have different needs from human beings. They deserve their own taxonomic category. However, there is no radical ontological schism between plants, animals, or humans. Plants are not zoocentrically dualised as inferior and are not placed at the bottom of a natural value-ordered hierarchy.

The autonomy of plants and their heterarchical relationship to us is recognized. Plants are regarded as kin, and are incorporated into general and specific kinship relationships—relationships of caring or solidarity, which are often "based on consubstantiality."[45] This approach to plants is coupled with a strong recognition that plants *are* different to human beings.[46] This difference is most strongly expressed in the act of predation. In a similar way to the ancient Indian material, Indigenous peoples recognize that the act of using plants is often an act of violence. Unlike in the biblical and Greek materials, recognition of the necessity of violence does not negate the recognition of personhood. Instead, the necessity of eating other-than-human persons is accepted, but like in Indian religions, ways are sought to mitigate the damage done to other beings.

As Chapter 5 discusses, to place plants in the ontological category of persons is neither fanciful nor deluded. The inclusion of plants in relationships of care is based upon close observation of plant life history and the recognition of shared attributes between all beings. Again, intent is paramount. This is a deliberate structuring of relationships in a heterarchy rather than a hierarchy. It is recognition of connectedness in the face of alterity—what Deborah Bird Rose has termed the "indigenous ethic of connection."[47] For plants at least, this contrasts sharply with what could be termed a Western ethic of exclusion.

One of the most important points in this text is that contrasting ways of understanding plant life can not be adequately split along easily demarcated lines. As has been noted, the case study of Buddhism shows that we must avoid constructing East-West dualisms. Similarly, Chapter 6 shows that an Indigenous-Western dualism is also flawed. An important component of this chapter is its recognition of the way in which contemporary European Pagans are also developing a more inclusive, more kinship based, less zoocentric relationship with plants. As seen in the Buddhism case study, this is another intentional turning away from zoocentrism, and has been inspired by engagement with Indigenous knowledges as well as ancient pagan sources.

Some of these ancient pagan sources are discussed in Chapter 6. The main argument in this chapter is that the fragmentary evidence from pre-Christian/Aristotelian Europe also depicts recognition of substantial kinship links between all beings in the biosphere. Using insights from contemporary animist scholarship, it is apparent that many pagan sources treat plants as fundamentally autonomous, volitional, communicative, relational beings. The notions of plant personhood and human-plant kinship are expressed in stories, poems, and myths. Common expressions of personhood and kinship are metamorphoses from human form to plant form. Unlike in the streams of thought that supplanted paganism, violence toward plants is acknowledged in several pagan texts. Chapter 6 puts forward the possibility that Western culture may have once have had a more appropriate way of relating to plants than that provided by zoocentric philosophies.

Perhaps the most interesting finding of this study is that recognition of many of the attributes of plant personhood and human-plant kinship is not restricted to the domain of religious studies. Chapter 7 argues that since the early nineteenth century, scientific evidence has steadily accrued which directly contradicts the hierarchy of nature. From Charles Darwin's early experimental work on the sensitive plant *Mimosa pudica* L., over a century of scientific observations contradict the notion that plants are passive, insensitive beings. Through nastic movements and tropic growth responses, plants have been shown for decades to display sensitive, purposeful, volitional behavior. Darwin's most important work *On the Origin of Species* also implicitly contains the idea that humans and plants are indeed related by descent.

Darwin's experimental work has also provided the platform for the development of the field of plant signalling. This area of plant science is outlined in Chapter 7 and is particularly interesting because it demonstrates that plants engage in abundant communication, both within their own bodies and with the beings in their environment. By evaluating plants on their own terms, it has also led to the development of the groundbreaking concept of *plant intelligence*. Plant intelligence's most vocal proponent, Tony Trewavas, argues that plants are increasingly being shown to demonstrate more sophisticated aspects of mentality such as reasoning and choice. Instead of displaying this through movement, plants differ from animals by using phenotypic plasticity to express behavior.

Another exciting development in contemporary scientific research is the accrual of evidence demonstrating that plants have the physiology to support sophisticated mental activity. As Darwin first discovered, there is increasing evidence that this intelligent behavior is directed by a multitude of brain like entities known as meristems. The work of František Baluška, Stefano Mancuso, and others in the nascent field of *plant neurobiology* is putting forward the notion that plants have sophisticated, decentralized neurosensory systems. Buried within contemporary plant science literature is a growing awareness that plant behavior has many of the hallmarks of mentality. Such pioneering scientific work in many ways echoes the recognition of the attributes of sentience and personhood that have long been pinpointed in Indian religious thought and Indigenous knowledge systems.

Chapter 7 takes a systems approach to matters of mind, avoiding Cartesian dualisms in order to describe how plants and humans share a basic, ontological reality as perceptive, aware, autonomous, self-governed, and intelligent beings.

Like other living beings, plants actively live and seek to flourish. They are self organized and self created as a result of interactions with their environment.[48]

The emergence of this evidence within a culture dominated by the findings of science adds great weight to the claim that our general perception and treatment of plants is both inaccurate and inappropriate. It also indicates the appropriateness of other philosophical traditions that relate to plants in inclusive, nonhierarchical, dialogical ways. In the words of Andrew Brennan, it provides "a context within which an attitude of care about natural things makes sense."[49]

The sceptic can of course ignore this accumulated knowledge and continue to exclude plants from moral consideration, but this option comes loaded with environmental consequences. Moreover, with an awareness that plants are autonomous subjects, continued instrumental exclusion must be viewed as deliberate disrespect. As Plumwood eloquently states, "We do them an injustice when we treat them as less than they are, destroy them without compunction, see them as nothing more than potential lumber, woodchips or fuel for our needs. . . ."[50]

Human-Plant Relationships of Care and Consideration?

With guidance from animistic cultures and the evidence from contemporary plant sciences, the latter stages of this study argues for recognizing plants as subjects deserving of respect as other-than-human persons. It advocates including plants within human ethical awareness with a view to Callicott's reminder that "an ethic is never perfectly realized on a collective scale and very rarely on an individual scale. An ethic constitutes, rather, an ideal of human behavior."[51] In the pages of this work, this ideal human behavior is grounded in a particular understanding of morality. Although there are many understandings of morality, most share the notion of right conduct toward others. In view of our Earthly kinship with both human and *other-than-human* persons and the interactions between these persons which allows life on Earth to thrive, discussions of ethics in this work are rooted in the recognition of these relationships.[52] Moral consideration in this respect is simply considering the flourishing of the other beings in our lives. In an ecological context, moral action is enacted respect and responsibility for the well-being of the others with whom we share the Earth.

The concluding chapter examines the implications of a new awareness of plant life and the development of a Western idea of plant personhood. Taking the findings of this study to their logical conclusion, the recognition of plants as autonomous, perceptive, intelligent beings must filter into our dealings with the plant world. Maintaining purely instrumental relationships with plants no longer fits the evidence that we have of plant attributes, characteristics, and life histories—and the interconnectedness of life on Earth. From another angle, conserving the natural environment is no longer sufficiently served by an anthropocentric account nor a zoocentric account of moral consideration. A stronger account of moral consideration centered on the other-than-human rather than human is needed in order to both evaluate and prevent the occurrence of "environmentally destructive human action that has little or no [immediate] negative effect on human beings."[53] In contrast with the focus of animal rights theory, in a biosphere dominated by plants, this turning toward the other-than-human cannot be at the implicit exclusion of plants from the class of morally considerable beings.

The concluding chapter discusses how this developed idea of plant personhood could become manifest in human moral behavior toward the plant kingdom and nature as a whole. Under the influence of Erazim Kohák, and the ethical theories of Zygmunt Bauman, the purpose of the concluding chapter is not to construct a list of proofs for moral consideration nor a system of ethical rules toward plants. Rather, its purpose is to discuss the possibilities for including plants within the realm of moral consideration; for the sake of individual plants and plant species and for those animals and humans whose lives depend on their survival.[54]

Purely instrumental relationships with plants are found to be ecologically destructive. The backgrounding of plants is dangerous because it severs opportunities for dialogical interaction between humans and the environments in which they live. Lacking in meaningful relationships of kinship, care, and solidarity, we risk complete human ecological dislocation. As Plumwood astutely observes, by distancing ourselves from the beings around us "we not only lose the ability to empathize and to see the non-human sphere in ethical terms, but also . . . get a false sense of our own character and location that includes an illusory sense of autonomy."[55]

By distancing ourselves from plants and denying their autonomy, we jeopardize a true sense of human identity, situatedness, and responsibility. Only in the company of others do we arrive at the true sense of our own personhood and ecological identity.[56] The risk we run by ignoring the personhood of plants is losing sight of the knowledge that we humans are dependent ecological beings. We risk the complete severance of our connections with the other beings in the natural world—a process which only serves to strengthen and deepen our capacity for destructive ecological behavior. This is humanity's worst type of violence.[57]

The concluding chapter also argues that one way to work toward restoring care-based human-plant relationships is through ecological restoration. With a revolutionised understanding of plants, restoring plant habitats can be a powerful and direct method for engaging in dialogue with plants as individuals, species, and communities. Here the idea of dialogue is based on the thinking of the scholar Mikhail Bakhtin.[58] Again in the terms of Bakhtin, one of the defining essences of dialogue is that "unlike monologue [it] is multivocal, that is, it is characterised by the presence of two distinct voices."[59]

Dialogue allows the recognition of the others "voice," standpoint, and needs.[60] One of the ways in which human beings can enact dialogue with plants is to give them the space they need to thrive and communicate with the world around them. In restoration, the needs of plants can be put first, and dialogue can ensue in the space that restoration creates. Because of the often fragile nature of restored ecosystems, restoring plant habitats is perhaps the best way of actively reestablishing personal care-based relationships with plants. In this way, restoration can be be viewed as a way of engaging in an active dialogue with plants, in which their voices come first.

1

THE ROOTS OF DISREGARD

Exclusion and Inclusion in Classical Greek Philosophy

> That we are a plant not of an earthly but of a heavenly growth,
> raises us from earth to our kindred who are in heaven.[1]
> —Plato

In the search for philosophies that have systematically backgrounded plants as passive, inert beings, while excluding them from moral consideration, our first stop is the work of classical Greece's most famous philosophers. Against a background of myth and story in ancient Greece (which is detailed in Chapter 6), in the fifth and sixth centuries BCE, a new, more critical way of understanding the world began to materialize.[2] Much has been made of the origins of purely scientific philosophical enquiry, but the change from an animistic, relational world to one based upon critical observation was a gradual process.[3]

The emergent pre-Socratic philosophy did not completely discard its Greek religious heritage. For example, the earliest pre-Socratic philosophers such as Thales still appear to have recognized the idea of an embodied divinity. Thales is famously reported to have exclaimed that all things are full of gods, Anaximenes of Miletus considered air to be a god, and for Xenophanes, God was the underlying substratum of the universe.[4] There is also recognition of substantial kinship between all things. For Thales, everything was made from water; for Anaximander the original stuff which linked everything in the world was air; while according to Heraclitus, this substance was fire.

Unfortunately there are few surviving references to plants by the pre-Socratics; however, the work of Empedocles provides a basis from which to assess the subsequent philosophical developments of Plato and the Peripatetics. One of the most noticeable features of Empedocles' treatment of plants is that there appears to be an explicit recognition of kinship between plants and other living beings. An examination of the surviving fragments of the *Physics* will serve to illustrate this kinship. In one passage from the *Physics,* Empedocles depicts an explicit substantial kinship between humans, animals, and plants by describing the "four roots" (earth, fire, water, and air) and how they serve to unite living beings:

> Trees sprang from them [the four roots], and men and women, animals and birds and water nourished fish, and long lived gods too, highest in honour.[5]

In the work of Empedocles, it seems that all these living beings have a common origin. Indeed it is noticeable that plant life is depicted as emerging before other living beings.[6] As in the animistic myths of ancient Greece that will be detailed in Chapter 6, Empedocles attributes aspects of sentience to all beings that emerge from the four roots:

> All things are fitted together and constructed out of these [the four roots], and by means of them they think and feel pleasure and pain.[7]

As for animals, the Empedoclean outlook is that a plant's thinking is most often directed toward the search of sufficient food, the acquisition of which is thought to cause pleasure, while the deficiency of food is thought to cause pain.[8] The recognition of this subjectivity in animals led Empedocles to famously decry animal slaughter for human consumption.[9] In the fragments of Empedocles' work, there is also a similar suggestion that plants should be treated with some respect. In the *Karthamoi*, Empedocles advises "to keep completely from leaves of laurel" in order to avoid harm and injury to them.[10]

In this respect, Empedocles retains strong elements of the ancient animistic understanding. In Hesiod's *Theogony*, it is the Earth that gives birth to everything, including all the gods, human beings, and the plants.[11] Thus, there is a sense of kinship between all beings. Greek mythology also presents kinship in the form of transformation and links this to the possession of person-like qualities of sentience. A common depiction of trees and plants involves them undergoing suffering in the same way as animals and human beings. Chapter 6 details the myth of the Heliades, daughters of the sun god Helios, who mourned for their brother so intensely that they were transformed into poplar trees. The poplars that emerged from the transformation demonstrate sentience by continuing to cry tears of pain and grief; a fact that demonstrates their ontological similarity to human beings as creatures capable of being harmed.

In Empedoclean thought, it is noticeable that there is a similar ontological connection between plants and human beings. Plants and humans are closely related—each made of the four roots and each displaying considered, thoughtful behavior. In Empedoclean philosophy, the recognition of this relatedness obligates human beings to consider the interests of plants in their behavior toward them.

However, although pre-Socratic philosophy retained elements of an animistic understanding, the developing philosophical discipline began to reject the myths and stories which personified other living beings and natural entities. Philosophers such as Xenophanes actively began to define their rationalism in opposition to mythology and Greek superstitious religious tradition.[12] The rejection of traditional poetry, story, and ritual began to be a rejection of the animistic nature of ancient Greek society. In this respect, although Empedocles urged respect for plants, the pre-Socratics movement also heralded a shift away from a society focussed on maintaining respectful relationships with nonhumans, toward the prioritization of rational, causal explanations of natural phenomena.[13]

Plato's Plant Philosophy

From this critical, more rational platform, Greek philosophical thought was to have a huge influence on Western perceptions of, and behavior toward, plants. The primary process that can be identified in the classical Greek philosophy is the establishment of plants as passive creatures with no capacity for intelligence or communication. This can be clearly seen in the philosophy of Plato (ca. 427–347 BCE), which marks a turning point in the Greek tradition. Although, like those of Empedocles, Plato's ideas appear to contain remnants of ancient, animistic Greek thought; they also demonstrate a turning away from plants being viewed as related, active, autonomous beings.

In her seminal ecofeminist critique, *Feminism and the Mastery of Nature*, Plumwood regards Platonic philosophy as one of the key Western sources for the "denial, exclusion, and devaluation of nature."[14] Plumwood writes:

> Platonic philosophy is organised around the hierarchical dualism of the sphere of reason over the sphere of nature, creating a fault line which runs through virtually every topic discussed, love, beauty, knowledge, art, education, ontology. . . . In each of these cases the lower side is that associated with nature, the body and the realm of becoming, as well as of the feminine, and the higher with the realm of reason.[15]

Plumwood discusses the formation and justification of this hierarchical dualism in depth, and the following sections owe a great deal to her scholarship.

However, for brevity, to elucidate the main principles, I will focus on discussing the philosophy of two types of causation, which in the *Timaeus* is dualized into a higher and lower form. The higher form is considered to be intelligent in nature, pertaining to a "rational principle or Form, and which is the true form of causation."[16] This type of causation is allied to the abstract, unchanging Platonic Forms, which are considered superior to ordinary, unchanging, material reality.[17] This rational causation is considered to be the source of all goodness.[18] In Platonic philosophy, it is contrasted with the lower causative principle, the material, which unlike the higher form is neither eternal nor unchanging. Plato regards this form of causation to be irrational, lacking in intelligence and ultimately disorderly.

By way of a very brief summary then, one of the principal reasons for reason's considered superiority is its alliance to the realm of the abstract, unchanging, divine Forms. As the material form of causation is not universal or eternal, it is generally regarded as inferior. Reason is also regarded as superior to nature, because it imposes order on the disorderly, insensate material realm.[19] As Plumwood makes clear, Plato attempts to both extend and justify this philosophy in his cosmology to counter the material philosophy of Democritus. In the cosmology of the *Timaeus,* God, representing logos or the rational ordered principle, imposes an order on the cosmos in flux "because he believed that order was in every way better than disorder."[20] Thus, as the ordering principle of the cosmos, logos is considered superior. It is also ranked higher as it is able to shape the malleable, changing material world, which is described in the *Timaeus* as "neutral plastic material" "upon which differing impressions are stamped" by rational forces.[21]

From reading the Platonic dialogues, it is clear that in order to sustain a dualism between reason and nature it is necessary for Plato to divest reason from the plant kingdom, which dominates the natural world. As reason is regarded as the superior faculty, and is defining of superior human life, any domination of nature requires the emphasis of differences between humans and plants. Based upon this need for separation and exclusion, plants are divested of reason in the same way as slaves and women.[22] Portrayed as passive, mute beings, plants are thus more easily dominated as mere resources for human endeavor.

Plants are attributed with the lowest level of Plato's tripartite soul. In the above account from the *Timaeus*, they are given the appetitive soul, but are denied the spirited soul (activity and volition) and the rational soul (intelligence and self control).[23] The beings most repeatedly associated with the appetitive soul are slaves, women, and children—beings at the very bottom of Plato's social hierarchy. As they are thought to lack reason and self control, plants (along with slaves, children, and women) are naturally ruled by those male human beings who exercise rationality.[24]

When compared with the philosophy of Empedocles, it is possible to detect in Plato's *Timaeus* the beginning of a process of exclusion, which depicts the ecologically dominant plant kingdom as passive and limited in awareness. This exclusion appears to be very much an act of intellectual violence perpetuated in part to sustain a (false) collective identity of humankind as well as justifying resource use.[25] In the description of creation, the *Timaeus* provides details about the genesis of the plant kingdom:

> They mingled a nature akin to that of man with other forms and perceptions, and thus created another kind of animal.[26] These are the trees and plants and seeds which have been improved by cultivation and are now domesticated among us; anciently there were only the wild kinds, which are older than the cultivated. For everything that partakes of life may be truly called a living being, and the animal of which we are now speaking partakes of the third kind of soul, which is said to be seated between the midriff and the navel, having no part in opinion or reason or mind, but only in feelings of pleasure and pain and the desires which accompany them. For this nature is always in a passive state, revolving in and about itself, repelling the motion from without and using its own, and accordingly is not endowed by nature with the power of observing or reflecting on its own concerns. Wherefore it lives and does not differ from a living being, but is fixed and rooted in the same spot, having no power of self-motion.[27]

Like Empedocles before him, in this passage, Plato appears to recognize continuity between life forms. He is explicit that the plants are made by the gods from a nature that is akin to that of human beings. Also, as for other types of life, the *Timaeus* describes plants as having experiences of desire as well as of pleasure and pain. Yet, Plato differs from Empedocles in his hierarchical value-ordering of the different types of being, a value-ordering which sanctions domination.

Therefore, within this suppressive hierarchical framework, the recognition of relatedness and the ability to be harmed requires no modification of human behavior toward lesser beings. As Carone points out, "Plato does not seem to find anything wrong with feeling pain as long as it can serve a greater, positive end."[28] Plato's *Republic* is a slave-based society, fundamentally hierarchically ordered, with the pain and work of the slaves sustaining the existence of the three classes of the polis.[29] The slave is constructed as lacking in reason, in the same way as nature is painted as a chaotic state requiring ordering.[30]

Although Plato admits some continuity between plants and human beings, his work instigates the process of removing connectivities between plants and humans. In the above passage from the *Timaeus*, Plato appears to define the *phuta* (plants) principally by their growth, as the singular *phuton* also means "that which has grown." In rejecting ideas that plants had such faculties as activity, self motion, and awareness without providing any evidence, Plato displays a fundamentally zoocentric philosophy. In Plato's observations, as plants are sessile beings, they are described as only capable of growth. In comparison with the incessantly active animals, plants are rendered as passive, inactive, and unminded.

Thus, in the description of creation in the *Timaeus,* the real purpose of plants is to not to live and blossom for themselves, but to provide animals and humans with food. Immediately after plants are painted as passive creatures, Plato writes that the "superior powers had created all these natures to be food for us."[31] Although it is clear that plants *are* food for animals, the fact that Plato portrays plants as unminded, inactive beings incapable of flourishing means that they can have no purpose of their own.

Aristotle's Hierarchy of Soul

Although we can trace the emergence of instrumentalism and hierarchical ordering in the *Timaeus,* the work of Aristotle (384–322 BCE) greatly expands and intensifies this process of exclusion. Building upon the ideas of Plato, Aristotle continues the drive toward separation and discontinuity, the drive that Plumwood considers to be characteristic of Western philosophy.[32] One of the most noticeable aspects of Aristotle's philosophy of the natural world is his hierarchical ordering of nature and the bracketing of plants into a lower class of being. This devaluing of plants is more explicit in the work Aristotle as it explicitly forms part of his famous theory of the soul. Breaking from the tradition of the pre-Socratics, Aristotle regards the soul as a form, rather than as a separate substance. The nature of the soul for Aristotle is set out in *De Anima*:

> Soul is substance as the form of a natural body which potentially has life and since this substance is actuality, soul will be the actuality of such a body.[33]

In this view, the soul is not just a form but a form that has life, "the ensouled is distinguished from the unsouled by its being alive."[34] All living things are thus attributed soul, while inanimate objects are not ensouled. Life and soul can be said to be interlinked. From this basis, as well as the soul being a form, Aristotle's concept of soul could also be interpreted as the principle and practice of life itself.

> For the plant will be torn apart if there is nothing that prevents this and if there is such a thing this will be the soul.[35]

It is clear from this passage that the soul of a being sustains life and prevents decay and death. In the thinking of Aristotle, the soul organizes and preserves life. It also directs a being's mode of living and sets limits on change and growth.[36] Such maintenance of integrity could lead to the conclusion that ensouled beings are autonomous and active. However, Aristotle's description in *De Anima* of different "faculties" or effectively *types* of soul, denies plants this nature. Aristotle describes the faculties of the soul as "nutritive, perceptive, desiderative, locomotive and intellective."[37] These faculties or different aspects of soul represent the different capacities of living beings. In the philosophical tradition, these five faculties of the soul have generally been divided into three levels.[38] The first of these is the nutritive level, which confers the ability to feed and reproduce. The second level is the sensitive and the third level of the soul is the intellectual.

The works of Aristotle recognize that all living beings possess the first of these levels, the nutritive or vegetative soul. The nutritive soul is regarded as the most basic in the sense that its existence is fundamental to the possession of the other faculties. The ability to gather food and reproduce underpins the abilities to sense and to reason. However, in *De Anima* as well as in *Parts of Animals* and *Nicomachean Ethics*, the higher faculties of the soul are not attributed to all beings. In the tripartite division of soul, plants are only attributed the ability to feed and reproduce. Describing the nutritive soul, Aristotle stresses:

> Now this faculty can be separated from the others but the others cannot be separated from this in mortal things. And this is obvious in the case of plants as they have no other potentiality of the soul.[39]

In this description of plant faculties, Aristotle extends the Platonic separation of plants and animals. Denied the ability to perceive their environment through the sense of touch, plants are perceived to be without any of the other faculties of soul. Unlike all other ensouled beings, plants are conceived as being without awareness and mentality. This explicit philosophy serves to counteract the idea that as ensouled beings they maintain their own integrity. Therefore, life processes, such as the feeding and reproduction of plants, are rendered as passive, mechanical processes. In this state, plants become truly vegetative, without any self-direction and unable to assess the environment around them:

> Plants having only the nutritive, other living beings both this and the perceptive soul. But if they have the perceptive faculty they also have that of desire. For desire is appetite, passion or wish, all animals have at least one of the senses (namely touch) and for that which there is perception there is also pleasure and pain and the pleasant and the painful

and for those for whom there are these there is also appetite, the desire for the pleasant. And they also have perception of their food.[40]

Based upon a denial of the obvious, that they are sensitive and perceptive, in the system of Aristotle, plants were also excluded from the highest faculty of the soul, the possession of intellect or mind. This is again expressed clearly in Book 1 of *Nicomachean Ethics*:

> For the vegetative element in no way shares in reason, but the appetitive and in general the desiring element in a sense shares in it, in so far as it listens to and obeys it.[41]

Again, Aristotle follows Plato's pronouncement in the *Timaeus* and authenticates it for the study of natural history. In this excerpt, Aristotle removes volition, choice, intelligence, and communication from the growth activities of plants, further asserting a subjective vacuum in the plant kingdom. In the botanical tradition, this is perhaps the earliest assertion that plants are divest of mental attributes, and all subsequent assertions can be said to draw upon it in some way.[42]

Unlike in Empedocles, and the founding myths of ancient Greece in which plants were recognized as ontologically related to animals and human beings, Aristotle's strict divisions remove the sense of a kin-based relationship between humans and plants. The voices of plants are completely silenced in Aristotle's work—reducing their lives to nothing more than feeding and reproduction.

As Aristotle attributed them only with a nutritive soul, the lowest of the three divisions, plants have traditionally been viewed as being at the bottom of a "psychic hierarchy" or *scala naturae*.[43] This value-ordering of the soul's faculties rests on several bases. Carone regards the nutritive soul as basic because it supports other faculties of soul, but can also exist as a stand alone faculty, as in plants.[44] However, this does not fully explain the value-ordering of the different souls. Why should the nutritive faculty be regarded as basic solely because it can function alone?

The rational soul is placed at the apex of existence in the natural world partly because the *Nicomachean Ethics* relates that the proper exercise of reason results in happiness, the highest human good:

> For we praise the rational principle of the continent man and of the incontinent, and the part of their soul that has reason, since it urges them aright and towards the best objects.[45]

It is reason that allegedly sets human beings apart from others, and is the faculty which in most senses defines human beings,

> Enough of this subject, however; let us leave the nutritive faculty alone, since it has by its nature no share in human excellence.[46]

Reason is valued above the nutritive faculty as Aristotle believes that reason allows human beings to be virtuous and achieve happiness. It is regarded as the element that prompts human beings into action and is valued above the passive nutritive faculty, "for in all cases that which acts is superior to that which is affected."[47] This attitude has parallels with a Platonic philosophy, which valued reason above nature on the basis that reason allowed the imposition of order. Ultimately however, it can be said that reason is valued above all because it is regarded as the hallmark of the human being. In other words, the higher faculties of soul are higher purely because they are thought to belong solely to human beings. This value-ordering is fundamentally anthropocentric, with humanity becoming the yardstick for value.[48] This anthropocentrism in Aristotle is perhaps most famously expressed in a passage from the *Politics*:

> In like manner we may infer that, after the birth of animals, plants exist for their sake, and that the other animals exist for the sake of man, the tame for use and food, the wild, if not all at least the greater part of them, for food, and for the provision of clothing and various instruments.[49]

This famous passage agrees with the *Timaeus* and clearly expresses the notion similar to that found in the Bible (see Chapter 3) that all the entities in the natural world exist for the sake of humanity. In this anthropocentric world, humanity is the apex of existence, for Aristotle regards all other beings to have been created for our use. This hierarchical notion of instrumental use clearly echoes Aristotle's hierarchy of the soul. In fact this hierarchy of use is entirely dependent on the tripartite division of the soul. The lowest beings, the nutritive plants, exist for the sake of animals and humans, while the sensitive animals exist for the sake of the rational humans. The important point here is that without rendering plants as passive creatures—inactive, mute, and insensitive—it is much more difficult to make such claims on the plant kingdom. In order to claim them solely as *instruments* for human use, Aristotle violates the autonomy of plants and animals, indeed strips them of any autonomy or subjectivity.

The refusal to acknowledge any aspect of agency, sensitivity, or mentality in plants appears to be a deliberate political ploy—in much the same way as Aristotle (and again as Plato before him) depicts slaves as naturally lower beings in order to justify their bondage.[50] In this context, the goal in backgrounding plants is to achieve the untrammeled use of plant resources. In Aristotle, therefore, it is not unreasonable to conclude that plants are backgrounded, rendered as passive and mute, in order to achieve human domination. The resulting instrumental relationships serve to nullify any notions of relatedness or responsibility of care toward plants for their own sake and so do away with inherent limits on human claims.[51] This slavery of the natural world also has very important ramifications for human beings, for as Vlastos notes, it is used to justify the domination and slavery extant in human society.[52]

Zoocentrism

As well as being founded on anthropocentrism, this exclusion of plants in Aristotle relies equally on a zoocentric appraisal of plant life.[53] For Aristotle to consign plants entirely to the nutritive soul, his observations of plants should have excluded the possibility that plants were able to perceive their environment. Yet although Aristotle's theory deemed that plants do not have perception, a passage in *De Anima* clearly demonstrates that Aristotle observed plants responding to external stimuli in a very similar way to animals:

> It is also clear why it is that plants do not perceive, *though they have a psychic part and are in some way affected by the touch objects*. After all they become hot and cold. The reason is that they do not have a means nor such a principle as can receive the forms of the sense objects.[54]

Aristotle clearly observed plants responding to both touch and to changes in ambient temperature. In the above passage, he even goes so far to admit that this constitutes a kind of active mentality in plants. In *De Anima*, Aristotle asserts that beings with the sense of touch are able to perceive their environment, for "those living things that have touch also have desire."[55] It is clear from the passage above that Aristotle recognizes that plants respond to touch in a similar way to other living beings. In Aristotle's system, plants *logically* deserve recognition as sensitive, perceptive beings.

Instead, plants are denied this, partly because Aristotle was unable to observe the means by which they could achieve perception. The zoocentric bias here is clear; because plants do not have the discernible brain tissue of most animals, Aristotle deems them incapable of perception. Rather than considering them on their own terms, instead the abilities of plants are defined from a zoological perspective—evaluated using the physiology and capacities of animals.

Thus the zoocentric bias which is discernible in Plato pervades Aristotle's philosophy of the natural world. This arises not from innate physiological biases as suggested by Hallé, Wandersee, and Schussler. In Aristotle, the *zoocentrism* is more akin to a philosophical stance, underpinned by limited observations and a general bias emerging from his scholarly orientation. In his studies of the natural world, Aristotle focuses his attention on animals, and as a prolific zoologist, Aristotle demonstrates a remarkable understanding of the structure and function of the anatomy of animals. One of the consequences of this academic specialization is that in his dealings with plants, Aristotle almost always defines and evaluates plants using *zoological* criteria and considers plant physiology as a *series of lacks*.[56] An example of the apparent simplicity of the plant kingdom is expressed in a comparison of plant and animal organs:

> Now this kind will include any body that has organs—and even the parts of plants are organs, though completely, as for instance the leaf is a covering for the pod and the pod for the fruit while the roots are like the mouth in that both draw in food.[57]

Although Aristotle is regarded as a great empiricist, he is not known as a particularly fine botanist. This flawed understanding of plant anatomy represents the general level of Aristotle's botanical investigations, and from this rendering of plant morphology, Aristotle evaluates the plant kingdom. Aristotle writes of three major differences between plant and animal physiologies. The first of these is the lack of a control system. The other differences between plants and animals are found within *Parts of Animals* Book 2 which states:

> As for plants, though they also are included by us among things that have life, yet are they without any part for the discharge of waste residue. For the food which they absorb from the ground is already concocted, and they give off instead [as its equivalent] their seeds and fruits.[58]

The presumed differences found in plants are the inability to process food and to excrete waste products. Again, these life processes are perceived entirely from the perspective of zoological anatomy and physiology. In particular, the passivity of plants is reinforced due to the presumption that they undertake no processing of their food. As the food they needed is presumed to be adequately concocted in the ground, there must be no activity involved.

Unlike animals, plants are sessile. Animals find their food and process it by movement. Therefore Aristotle's apparent reasoning is that as plants are sessile, they do not accomplish such food acquisition and processing. The roots of the plant are compared to the mouths of animals and are presumed to act as a passive receptacle for the uptake of food. As this concocted food in the ground is thought to be just what plants need, it is presumed that they do not give off waste products.[59] Instead, plants are conveniently thought to give off only seeds and fruits, solidifying their role as instruments for human and animal consumption.

The Father of Botany

Theophrastus (371–ca. 287 BCE) is a colossus of the Western botanical tradition and is commonly bestowed the title the Father of Botany.[60] As the head of the Lyceum, Theophrastus continued the work of Aristotle, and it is thought that the task of researching the botanical element of natural history was given to him by his teacher. In addition to his numerous other works, it is presumed that botany was too large a labor for Aristotle to tackle properly.[61] In his two outstanding botanical treatises, *Historia Plantarum* and *De Causis Plantarum*, Theophrastus reported many of the findings that are now fundamental to his discipline. Theophrastus's works differ from those of Aristotle in their detailed observations of the plant world. Often, the intricate detail of plants' internal and external anatomy is noted with accuracy. For instance the different textures of the wood in different tree species, the complexities in root systems and various methods of germination were all observed.

Most of Theophrastus's botanical work is concerned either with systematic, physiological, or agricultural matters and is not (as was Aristotle's *De Anima*) geared toward an explicit philosophical analysis of the nature or capacities of plants. Theophrastus did write his own version of *On the Soul,* but unfortunately this text has not come down to us. The only surviving fragments of this work are quotations from Priscian and Themistius. Nevertheless, despite not being explicitly metaphysical, the erudite descriptions in *Historia Plantarum* and *De Causis Plantarum* contain implicit references that allow an understanding of Theophrastus's conception of plants.

Plants and Animals, Connectivities and Differences

> Your plant is a thing various and manifold; and so it is difficult to describe in general terms.[62]

In his *Metaphysics,* Theophrastus writes that to be known, each thing must be approached in the appropriate manner; "The starting-point and the main thing is the appropriate manner."[63] For botany, this involves careful and detailed observations of living plants. In his seminal botanical studies, he fully recognizes the differences that plants display. However Theophrastus does not find it appropriate to evaluate plants from a zoological perspective. Instead, the differences inherent in plant life suggest the need to deal with plants on their own terms, and so Theophrastus notes in the first few pages of *Historia Plantarum*:

> Perhaps we should not expect to find in plants a complete correspondence with animals in regard to those [parts] which concern reproduction any more than in other respects.[64]

Theophrastus's basis for understanding plants is to recognize that their lives are constructed in different ways to those of animals, a fact which implicitly rejects the zoocentric evaluation of Aristotle. For example, Theophrastus recognizes that the local environment is more important for an understanding of plants than of animals. As the next section explores, he was aware that plants are perhaps more influenced by their environments than animals. This is because a plant is "united to the ground and not free from it like animals."[65] As well as in reproduction, other major differences between plants and animals are recognized. In particular, Theophrastus pays attention to the incomplete correspondence between plant and animal growth:

> The number of parts is indeterminate [in plants]; for a plant has the power of growth in all its parts; in as much as it has life in all its parts.[66]

As well as being indeterminate in their growth, the number of plant parts is described as being constantly in motion. The recognition of this detail of plant anatomy is greatly significant because here Theophrastus is describing attributes in plants that are *lacking* in animals:

> It may be however that, not only these things but the world of plants generally, exhibits also other differences as compared with animals, for as we have said the world of plants is manifold.[67]

Reading this passage from *Historia Plantarum*, it is hard not to detect Theophrastus's admiration and wonder when faced with the plant kingdom. Certainly the "manifold" plant world comes across as more diverse, unusual, and intricate than the world of animals. In this manifold world, plants are implicitly acknowledged as possessing many different qualities and capabilities not necessarily found in their animal kin. Yet it is characteristic of Theophrastus that this difference is stated without reference to any supposed inferiority on the part of plants. Unlike Plato and Aristotle, it appears that he recognizes differences in plants, but does not seek to arrange these differences in a hierarchy of value with reference to anthropocentric or zoocentric criteria.[68] Theophrastus also worked toward finding similarities between beings, a stance which is confirmed by a passage in his *Metaphysics*:

> And in general the task of a science is to distinguish what is the same in a plurality of things.[69]

Thus, throughout the texts of *Historia Plantarum* and *De Causis Plantarum*, Theophrastus emphasizes connections between the anatomy and life history of plants and animals. In the search for these connections, the shedding of plant parts such as leaves in winter, and the shedding of animal parts, like horns, are regarded as analogous processes. The process of decay and death in plants and animals is described as being due to the same event, that of the failing of the moisture and warmth that belong to both.[70] Theophrastus echoes the mythology of the ancient Greeks (detailed in Chapter 6) and equates the sap of plants with the blood of animals. As well as their implict kinship in shared blood, other parts of plants are connected to those of animals. The bark, wood, and core tissue of plants are described as being made up of "sap, fibre, veins, flesh."[71] Theophrastus also takes the analogy further, labelling the core of the tree as the "heart" or "heart-wood," an expression still in use today.[72]

In *Historia Plantarum*, he is explicit that such comparisons exist only as an aid to understanding for people more familiar with the animal world.[73] They are a bridge to understanding and have the added function of showing and emphasizing the connections and kinship between animals and plants. They are not an attempt to define plants from a zoological perspective.

Historically it is important that Theophrastus took great pains to clarify that these names for the parts of plants are borrowed from animals due to a superficial resemblance. They were an aid to understanding and a means of bridging the gap between the plant and animal. They were never intended to be taken as literal explanations of plant structure and function because for Theophrastus the world of plants deserved evaluation on its own terms. However, it must be noted that an unfortunate historical consequence of this attempt to explain plants in terms of animal parts, was that plants continued to be commonly regarded in reference to animals. As Chapter 2 details, thinkers such as Pliny, who relied heavily on Theophrastus, did not understand the importance of Theophrastus's warnings and were less careful about *judging* plants from within this animal framework.

Plant Autonomy

The fundamental recognition that plants should be treated on their own terms manifests itself in Theophrastus's recognition of the independence and autonomy of the plant kingdom. In contrast to the backgrounding philosophies of Plato and Aristotle, through his careful description of fruit formation in *De Causis Plantarum*, Theophrastus recognizes a purposeful autonomy.[74] The first important description of fruit anatomy reads:

> There is to be sure a concoction of the pericarpion [fleshy pericarp of a true fruit], but there is another of the fruit proper [the seed], and the former concoction—serves to provide man with food, the latter serves

the generation and perpetuation of the tree, *this being what fruit and seed are for*. Each of the two concoctions interferes in a way with each other; with greater fluidity and size in the pericarpion goes smaller fruit [seed] and with larger fruit [seed] goes a smaller, harder and more ill flavoured pericarpion.[75]

While the detailed understanding of fruit anatomy is remarkable in itself, what makes this passage even more outstanding is the implicit recognition that the fruit and seed are not *designed* for man or animal to eat. There is a clear distinction between the proper seed that reproduces the plant, and the fleshy part of the fruit which is appropriated by man. Theophrastus recognizes that from a botanical point of view, one could say from the plant's point of view, the fruit and the seed are not aimed at satisfying man. The fruit and seed are for the plant to continue its lineage.

A sceptic could argue that the continuation of the species is very much in the interest of man and, thus, could quite easily be part of a system that was ultimately created for the satisfaction of human needs. However, if we also consider Theophrastus's studies on the different relative sizes of wild and cultivated fruits (and seeds), the only reasonable conclusion is that plants are considered to be autonomous beings with their own purpose.[76] Plants are not solely on Earth to feed human beings:

> Of the two ripenings this of the seed is the more important for reproduction, that of the pericarpion the more important for human requirements. To which of the two ripenings we are to assign the greater achievement by the tree of its goal is another question. Indeed if we assign it to the ripening of the pericarpion we should have to say that in plants whose leaves (or again whose roots) we use alone, as vegetables, the concoction of these parts is the more important [for humans]; and yet the goal lies here in their seeds, which we do not use for food at all.[77]

Wild plants are described as having much larger seeds than cultivated plants. By careful observation of a plant's fruits it is fairly obvious that free-living plants show marked preference to flourishing and continuing their own species, rather than producing food for human beings. In these two short passages, Theophrastus's studies lead to the recognition that plants have their own goals in life. A passage from *De Causis Plantarum* helps explain the Theophrastean understanding of what those goals might be:

> The nature [of a plant] instead always sets out to achieve what is best, and about this (one may say) there is agreement.[78]

In this instance, "what is best" refers to what is good for a plant, or its final cause. Rather than his *extrinsic* teleology that plants are created for the purpose of feeding humans, Theophrastus follows the *intrinsic* aspect of Aristotelian teleology, that there is a good to be had for an organism and that this good is a final cause to be striven for. Achieving this final cause is the purpose of the plant, not satisfying the needs and wants of human beings. With the establishment of *intrinsic* teleology Theophrastus renders the growth and reproduction of the plant *for itself*, as the cause to which its life is directed. This is acknowledged with the line "yet the goal lies here in their seeds, which we do not use for food at all."[79]

Significantly the recognition of plants as autonomous beings, rather than as slaves for humankind, is accompanied by descriptions of the striving, intentional, perceptive behavior they exhibit in order to fulfil their own purposes. To achieve what "is best," the plant must be able to know what is best, and so their flourishing in both current and future generations is described as being achieved via intentional, discriminatory activity. The descriptions of intentional, perceptive plants in Theophrastus are incredibly important as they are the first, albeit implicit, recognition of mental phenomena in plants in the Western botanical tradition.

Plant Action and Perception

Conduct and activities we do not find in them, as we do in animals.[80]

In light of Aristotle's work, this passage could be read as a support for his hierarchy of value. Some authors have indeed done this and claimed that Theophrastus went even further than Aristotle by removing any aspect of soul in plants.[81] However, in Theophrastus's extant writings there is no explicit doctrine of the capabilities of plants and no explicit support for the imprisonment of plants within the bounds of the nutritive soul. In light of his more vocal pleas to consider plants on their own terms, and the recognition of plant autonomy, it appears that the above passage is an attempt at a fair appraisal of the differences in plant and animal life. Just as he recorded the ability of plants to grow from different parts, so he notes that they are unable to move around and behave in the same way as animals.

The disagreement with Aristotle that plants are not passive beings can be seen in Theophrastus's understanding of sense perception. In his commentary on Theophrastus's *On the Soul*, Priscian asserts that Theophrastus did not think it possible for living beings to be devoid of sensory awareness. For Theophrastus, the possession of sense is "completive of the living thing."[82] Unlike in *De Anima*, being alive in Theophrastean thought entails being aware of the environment.

A passage from *Historia Plantarum* has parallels with Aristotle's recognition that plants are sensitive to changes in ambient temperature. However, the Theophrastean understanding of perception leads to startlingly different conclusions. During a description of how environmental conditions affect plant growth, Theophrastus remarks:

> The roots of all plants seem to grow earlier than parts above ground (for growth does not take place downwards). But no root goes down further than the sun reaches, since it is the heat which *induces* growth.[83]

Within this description of the patterns of root growth, there is a clear acknowledgement of perceptive and receptive capabilities in addition to the basic recognition of growth.[84] This passage strongly suggests that the plant is able to perceive the presence of heat and gear its growth to this change in its niche. Aristotle describes the same phenomena in *De Anima*, but he does not admit this as an example of perception, because the plants have no discernible tissue that receive the form of the sensory object. In contrast, Theophrastus succinctly describes the ability to sense the environment and to respond with active, directed growth.

Much of Theophrastus's work was geared toward agriculture, and he was one of the first Greek writers to record basic agricultural techniques.[85] Historically, one of the most significant agricultural descriptions was a record of the technique for pollinating the date palm.[86] In *De Causis Plantarum*, the date palm is also used as an example of trees that are not only able to sense their environment, but which actually *enjoy* and *prefer* certain localities over others:

> For some trees, delight in the one or the other excess [of heat], some favoring heat, like the date palm, others cold, like the ivy and silver fir.[87]

> But different trees differ in the degree. Some love moisture and manure, some not so much as the cypress, which is fond neither of manure nor of water.[88]

These descriptions of the likes and dislikes of plants are just two succinct examples of a way of writing about plants that pervades the work of Theophrastus. Trees are attributed with having *preferences* as to which environments they may grow in. They have the *enjoyment* of thriving in their preferred environment and only in the most suited environment will a plant be able to flourish.[89]

In the work of Theophrastus, there is recognition that trees are very much affected by the soils in which they grow, by the amount of shade and sun, and by strength of the winds in the area.[90] Underpinning this is the recognition that plants possess the ability to sense the differences in their localities and respond accordingly. This adjustment of the plant's nature to the prevailing environmental conditions is explicitly stated in *De Causis Plantarum*:

> Every plant must possess a certain adjustment to the season, since the season turns out to be more responsible than anything else. For all are seen to await their own appropriate season.[91]

From his close field observations, Theophrastus clearly recognized that plants had capacities which far outstripped the faculties of the nutritive soul. These were faculties by which the nutritive and reproductive capacities of the soul were *accomplished* but which extended beyond feeding, growth, and reproduction.

Linked in closely with the demonstration of preferences are the faculties of perception and awareness. In order to express preference, plants must be able to differentiate environments and act according to the prevailing conditions. Perhaps of even greater interest is the fact that as possessors of this desiderative nature, plants are also closely connected with mentality. In the *Metaphysics*, Theophrastus asserts that it is "the thinking faculty, in which indeed the desire originates."[92] Thus, as beings that clearly act out their likes and dislikes, it is not unreasonable to suggest that for Theophrastus, plants were also beings with the capacity of intellect.

Autonomy and Agriculture

The recognition that plants are autonomous led Theophrastus to question whether cultivating plants was a natural process. Theophrastus also displays a remarkable affiliation with ancient animisms (see Chapter 6) by acknowledging the harm done to plants by human beings. In *De Causis Plantarum*, he notes that one of the costs to cultivation for the plants involved is a shorter life. Theophrastus observed that cultivated plants expend their energy on forming the part of their anatomy that is more desirable to humans, rather than on their own overall growth and repair.[93] Thus, cultivation diminishes the overall flourishing of plants, by reducing their physical health and their capacity for reproduction. Like Plato and Empedocles, Theophrastus asserts that human activity could cause *suffering* to trees, something explicitly denied by Aristotle. In cultivated trees he remarks:

> Not only do trees that have borne to excess fare thus [live shorter lives] but even when they have borne a large crop trees *suffer* and often perish from depletion.[94]

In fact, concepts of plant sentience seem to have still been quite widespread in Greece at the time. Theophrastus describes the practice of driving iron pegs into almond trees, which have too much foliage and too little fruit; and records that "some call this 'punishing the tree,' since its luxuriance is thus chastened."[95] Unlike in Aristotle, it appears that the men doing this work acknowledge the harm in their acts.

Interestingly, the process of cultivation is thought not to be of complete detriment to the plants involved. Some of the positive points to cultivation (from the plant's point of view) are that plants receive plentiful food and water, which may be lacking in the wild. They are also kept free from competition and disease, which impede growth and reproduction. Theophrastus muses that such benefits could be a price worth paying for the harm experienced by the cultivated plant. In this way, the morphological changes brought about by plant breeding could also be viewed as appropriate and reciprocal, perhaps of benefit to the plant involved:

> If the nature of the plant *demands* that external aid for the achievement of what is better, it would also accept these internal modifications as appropriate to itself; and it is reasonable that it should *demand* and *seek* them.[96]

Rather than being a form of bondage, Theophrastus understands cultivation to be a collaborative, mutualistic, relationship between plants and humans. With this, Theophrastus envisages a more respectful form of farming, in which the cultivator engages in a partnership based on respect for the awareness and autonomy of the cultivated.[97] In a similar way to a kinship relationship, this invokes the need for reciprocal care and responsibility.

This Theophrastean understanding stands in stark contrast to the *scala naturae* of Aristotle in which the plant (and consequently much of nature) is rendered as passive and subservient to human beings. Thus, at the dawn of the natural sciences, we can pinpoint two very different understandings of plants emerging from the classical Greek tradition. One was based upon *exclusion, separation*, and *superiority*; the other was based more upon *inclusion, connection*, and appreciation of *autonomy*. As Hallé has pointed out, our contemporary understanding of plants resembles the first position far more than the second.[98] Yet rather than being a natural phenomenon, it is clear from this chapter that our perception of plants depends heavily upon our philosophical orientation. The following chapter builds upon this understanding and explores how the two different Peripatetic perceptions of plants have had contrasting influences on botanical studies. It aims to demonstrate that the ready acceptance of the Aristotelian zoocentric dualism and anthropocentric hierarchy has hugely influenced Western understandings of plants.

2

DOGMA AND DOMINATION

Keeping Plants at a Distance

[Man] is master of the vegetable tribes, which, by his industry he can, at pleasure, augment or diminish, multiply or destroy. He reigns over the animal creation; because like them, he is not only endowed with sentiment and the power of motion, but because he thinks.
—George-Louis Leclerc, Comte de Buffon[1]

By examining the work of some of botany's most important thinkers, it is possible to demonstrate that the authority of the Aristotelian view of plants has been instrumental in maintaining the position of plant life as inferior. Reading the work of botanists from Pliny to Linnaeus, it is clear that descriptions of plants as passive and insensitive are nearly always connected with human denial of their autonomy and the total appropriation of their purpose. This chapter argues that key figures in the history of botany have contributed to entrenching the view of plants as passive, mute, and morally inconsiderable. They have done this by dogmatically adhering to the work of Aristotle instead of engaging with their own observations on the nature of plants. It is clear that the continued perpetuation this philosophy has served to establish a *default position* of exclusion in which plants are commonly understood to be passive, insensitive, and unminded. Despite its basis in anatomical fact, the Theophrastean view of plant ontology has been repeatedly ignored in the botanical tradition, partly due to historical factors.

The proliferation of the Aristotelian position in Western botany is important, for as Herman Boerhaave says "botany is that branch of natural science which enables us, most happily and with the least trouble, to know plants and to remember them."[2] On the authority of the world's most influential botanists, this default position of plant exclusion has also been embedded into the philosophies of some of Western society's most important thinkers such as Bacon, Descartes, and Locke. While environmental philosophers have highlighted their work as being particularly influential in developing a destructive Western attitude toward the natural world, the botanical influence on their philosophies has been overlooked.[3] This chapter explores the Aristotelian basis for their perception of the plant kingdom, and argues that it is a significant contributor toward their ideas of *nature* as an inert, passive entity, available for subduing and dominating.

The Natural History of Pliny

After the cultural highs of the Ionian ages, instead of the more philosophical and perceptive work of Theophrastus, in the more materially orientated societies which followed, it was his practical work on medicinal plants that was of greatest influence. His compilation of descriptions on medicinal plants in the ninth book of *De Historia Plantarum* was used by Dioscorides in perhaps the most famous treatise of medicinal plants, *De Materia Medica*. With the loss of the Greek texts, until the Renaissance, in addition to *De Materia Medica*, Pliny the Elder's *Natural History* also became an authoritative work on botany. In the words of classical scholar John Healy, the *Natural History* was so heavily relied upon it "became a substitute for a general education."[4]

Pliny was one of the last classical scholars to draw upon the work of Theophrastus before it was lost to science for over a thousand years. Therefore, Pliny is an important starting point for exploring the work of key scholars in shaping the perception of plants in botanical science. Pliny was not only authoritative for a thousand years, but his perception of plants continued to influence important scientists such as Francis Bacon and John Ray, extending his influence well into the seventeenth century. Exploring the work of Pliny enables a discussion of the influences that have been important in developing the perception of plants in the botanical sciences. In this chapter, it provides a starting point for a concise historical analysis of the processes that have shaped Western perceptions of plants.

Pliny drew heavily on *De Historia Plantarum*, but in his discussion of the findings of Theophrastus, he gave the work his own interpretation. A brief reading of *Natural History* shows how the botanical discipline and the Theophrastean understanding of plants was shaped in his hands. While *Natural History* covers the entire natural world, Books 12 to 27 are almost wholly devoted to plants. As in the works of Theophrastus, Pliny focuses strongly on the uses of plants for agriculture and medicine. Although Pliny was influenced by Theophrastus, it is clear from the *Natural History*, that he was a much less careful scholar.[5] The following is a passage from the *Natural History* where Pliny compares the anatomy of trees and animals:

> And in general the bodies of the trees, as of other living things, have in them skin, blood, flesh, sinews, veins, bones and marrow. The bark serves as the skin. . . . Next to the bark most trees have a layer of fatty substance. Under this fat is the flesh of the tree and under the flesh the bones, that is the best part of the timber.[6]

This passage echoes the *Historia Plantarum*, but it is clear that the precision of observation and the careful caveats set in place by Theophrastus were casually discarded by Pliny. The lack of original, critical observation is particularly important for it led to Pliny creating fantastical anatomical features from his own imagination. As the following sections will show, the lack of proper engagement with living plants and the muddled copying of previous scholarship is an important factor in the development of a zoocentric, hierarchical attitude toward the plant kingdom.

In this instance, the unthinking copying of Theophrastus's work led to the descriptive anatomical terms borrowed from animals, which Theophrastus intended only as a guide to understanding plants more thoroughly—*defining* plants from a zoological perspective. While some argue that this zoocentrism is inevitable; a critical reading of this single passage from Pliny makes it clear that this zoocentrism in part arises from the subsumation of the differences between animals and plants. It also involves the placement of plants into an analytical framework in which animals are *prior* and implicitly superior. Such treatment of course harks back to Aristotle.

The harm that can arise from unthinking, dogmatic acceptance and copying of earlier works is exemplified by Pliny's rendering of Theophrastus's notion that plants exhibit purposeful activity. In the *Natural History*, Pliny makes implicit references to plants that exhibit a subjective nature. However, whereas Theophrastus's descriptions of plant subjectivity were based upon careful observation, Pliny's writing emanated from his limited interpretation of the sources he relied upon.

On one hand, the *Natural History* recognizes that plants have preferences for certain environmental conditions, such as weather and soils; Pliny describes the cypress liking a warm climate and having a "great dislike for snow."[7] However on the other, his descriptions of plant subjectivity are drawn almost entirely from his imagination. They are symbolic representations rather than actual descriptions of plant behavior. In places, trees are describes as having "wicked conduct," and Pliny records that "some plants like having a distant view of the sea."[8] Similarly, Pliny writes of the suffering of plants. Trees were described as being exhausted by bearing fruit. Pliny notes that trees are often found to be "famished with hunger" especially in winter time:

> Indeed if several years in succession should take this course even the trees themselves may die, since no one can doubt the punishment they suffer from putting forth their strength when in a hungry condition.[9]

This demonstrates an understanding, undoubtedly drawn from Theophrastus, that plants are perceptive beings, capable of being harmed. Despite this depiction, Pliny did not attribute autonomy to the plant kingdom. In effect, Pliny either missed, or chose to ignore, one of Theophrastus's most important findings—that the primary purpose of plants (from a plant point of view) is not to provide food and medicine for human beings, but to live and grow for themselves.

Pliny was heavily influenced by the Stoics and, although botanically reliant on Theophrastus, appears to also aver the Stoic philosophy that the world was ordered to serve a uniquely rational humankind. Echoing Cicero's assertion that the Earth was "designed for those only who make use of it," the *Natural History* contains many an instrumental claim on the plant kingdom[10]:

> Plants that she [Nature] engenders for the health or the gratification of men.[11]

> For it is for the sake of their timber that Nature has created the rest of the trees.[12]

> [Nature] put remedies even into plants that we dislike . . . and she had given already the soft plants I spoke of that make pleasant foods.[13]

In this collection of passages, it is clear that Pliny regards plants as useful objects, rather than as subjects. An interesting observation here is that each of the passages relates that it is *nature* and not God (or a godly creator) that has created the plants on Earth expressly for human use. Whereas several environmental thinkers have linked destructive ecological behavior primarily to the "loss of" a conception of *nature* as organic, purposeful, and minded—these passages from Pliny suggest that we need to pay more attention to the *constituents* of the natural world.[14] Pliny clearly recognizes an active, dynamic, directed nature—but at the same time still regards relationships with plants to be wholly instrumental—relationships that are actually instigated by nature. This instrumentalisation is part of the separation of humans and nature.[15] Instead of focussing on revivifying a homogenous, somewhat abstract nature, these passages from Pliny highlight the importance of establishing corporeal plants as autonomous, subjective individuals.

Aristotle and the Heart of the Vegetative Soul

After their misappropriation by Pliny, during the height of the Roman Empire, the works of Theophrastus ceased to be copied and translated. Sadly, they were lost to European thought from the end of the second century CE.[16] During what is often known as the "dark ages of botany" (ca. 200–ca. 1500 CE) only plants suitable for pharmacology were studied.[17] Other philosophical, systematic, and physiological ideas disappeared almost completely from botanical study. Rather than conducting original observations, a driving force was the strong tendency to rely on the authoritative texts of Dioscorides and Pliny for information about plants. Plagiarism of classical texts was symptomatic of the period, and because of it, the perception of plants ever narrowed.[18] The flawed authoritative influences of Dioscorides and Pliny served to entrench in the European mind during the Middle Ages the philosophy of plants as useful, subservient objects. [19]

Theophrastus's works were only recovered after original Greek manuscripts were found in the Vatican during the 1400s. Pope Nicholas V had them translated into Latin by Teodoro Gaza (ca. 1430 CE), and the botanists of the Renaissance were able to draw great inspiration and influence from them. The reacquaintance with Theophrastus had a great impact on the technical aspects of plant taxonomy and systematics, which had fallen into disarray during the Middle Ages. However, after a thousand year period—although the texts of Theophrastus were resurrected—his ideas of plant autonomy were buried under the dogmatic insistence that plants were inferior and subservient to human beings.

In sharp contrast, the philosophy of Aristotle flourished during the medieval period. An example of the authority of Aristotle is provided by considering the works of the great Christian scholar, Thomas Aquinas (1225–1274 CE). Aquinas was heavily influenced by Aristotle, and his *Summa Theologica* paid heed to Aristotle's tripartite division in the capabilities of living things.[20] As Chapter 3 details, Aquinas succeeded in embedding Aristotle's ideas into church doctrine, and thus, the idea that plants were without sense, movement, or intelligence was common fare in medieval Europe. With biblical sanction, the Platonic-Aristotelian dogma that plants were designed expressly for the use of animals and humans was also readily accepted:

> Hence it is that just as in the generation of a man there is first a living thing, then an animal, and lastly a man, so too things, like the plants,

which merely have life, are all alike for animals, and all animals are for man. Wherefore it is not unlawful if man uses plants for the good of animals, and animals for the good of man, as the Philosopher states.[21]

So, the authority of divine scripture (plus Aristotle), led Aquinas to accept the theory of natural slavery in which the passivity of plants was a sign from God that they were "naturally enslaved and accommodated to the uses of others."[22] The inability (or the unwillingness) to recognize plant autonomy and purposeful behavior was fundamental to their domination.

With the backing of influential theologians, Aristotelianism had a profound influence on huge swathes of medieval thought, and continued to be influential in the Renaissance. In botany, it also had a profound influence on the work of Andrea Cesalpino (1519–1603 CE), who is regarded as "unquestionably the greatest botanist of his century."[23] He is renowned in systematics for being the first botanist to attempt a natural classification of plants, placing great emphasis on the use of the reproductive parts for classification, a distinction that still holds today. Notably, Cesalpino was not just a botanist; he also taught philosophy, and like many Renaissance thinkers, he was heavily influenced by Aristotle.[24] Cesalpino's major work, *De Plantis Libris*, wholeheartedly reiterates Aristotle's denial of plant sentience:

> As the nature of plants possesses only that kind of soul by which they are nourished, grow and produce their like and they are therefore without sensation and motion in which the nature of animals consists, plants have accordingly need of a much smaller apparatus of organs than animals.[25]

Cesalpino describes animals and humans as possessors of a sentient nature because their mobile forms require it. Viewed from this zoological perspective, as sessile beings, plants lack the need for awareness, perception, and reason. This over reliance on Aristotle's zoocentric appraisal of plant life is remarked upon by the botanical historian Julius Sachs, who considers that under the influence of Aristotle, Cesalpino conceived plants as "an imperfect imitation of the animal kingdom."[26]

Zoocentrism is also influential in one of Cesalpino's most famous contributions to botanical science—his identification of an organ in plants which he postulated contained the nutritive soul. Cesalpino identified the point at which the root meets the shoot as the place of the plant soul and named this *cor plantarum*, or *cor* (heart). This *cor* was analogous to the heart in animals, which Aristotle regarded as the seat of the sentient soul.[27] It is a feature of the dogmatic acceptance of zoocentrism that in order to make plants fit into a zoological framework, early botanists such as Pliny and later Cesalpino were forced to propose plant organs for which there was no anatomical evidence. The fact that these anatomical oddities were easily dismissed led future thinkers to doubt the existence of the nutritive soul altogether, which further distanced plants from the categories of active, purposeful, autonomous beings.

Baconic Hierarchies

Traditional histories of science recount that Aristotelian scholasticism was of great influence during the Renaissance up until the rise of an objective science in the late sixteenth and early seventeenth centuries From this point onwards—in the minds of Europe's leading scientists—the deductive reasoning of the scholastics was supposed to have been supplanted by the objective evaluation of empirical evidence. In the current botanical context, this assertion requires greater nuance. This claim is based on the fact that it is possible to trace the uncritical acceptance of Aristotle's dogma of hierarchy and domination affecting the attitudes of some of the most influential minds of the scientific revolution.

The twin themes of hierarchy and domination are most easily identified in the thinking of Francis Bacon (1561–1626 CE), pioneer of scientific methodology. During his studies at Cambridge, Bacon became aware of the restrictions of the deductive, Aristotelian curriculum and as a result began to develop his famous theories of experimentation and observation. These were published mainly in his *Novum Organum* (1620) and have profoundly and permanently influenced the scope and practice of science ever since. However, despite his rejection of Aristotle's deductive reasoning, Bacon retained the zoocentric attitude of domination toward the plant kingdom.

At the time of his death, Bacon was working on the manuscript for his *Sylva Sylvarum, or A Naturall Historie In Ten Centuries*, his mature treatment of natural history published posthumously in 1626. This text addressed a range of topics spanning physics, medicine, alchemy, and natural history. In the sections that deal with plants, Bacon openly displays a predominantly instrumental attitude toward plants as "they are of excellent and general use, for food, medicine and a number of medicinal arts."[28] The influence of Aristotle is less obvious and stems primarily from Bacon's reliance upon the ideas of Renaissance writer Giambattista Della Porta (1535–1615 CE) and his text *Natural Magick*.[29]

Della Porta himself drew heavily on both the rediscovered texts of Theophrastus and Pliny's *Natural History*, to which he frequently referred directly.[30] However, his description of a plant's capabilities reveals that he also uncritically accepted the Aristotelian doctrine of human superiority and plant passivity. In his attempt to uncover the secrets of nature for the use of mankind, Della Porta describes man as "more excellent than other living creatures" because man alone possesses reason.[31] It is likely that he did this without criticism because it served his requirements, constructing the natural world as naturally subservient to mankind. Therefore in *Natural Magick*, Della Porta simply repeated the Aristotelian assertions that plants "have neither sense nor reason, but do only grow, are said to live only in this respect, that they have this vegetative soul."[32]

Like Pliny before him, Della Porta picked up some of the influence of Theophrastus's writing about the striving of plants, although he confused them with the Empedoclean theories of love and strife.[33] Seemingly also from his observations of his local flora, Della Porta gathered the impression that plants were not entirely passive. Although somewhat poetic, the *Natural Magick* includes descriptions of animated plants. For Della Porta, the vine *shuns* coleworts, and is *an enemy* of *Cyclamen*. Ivy is the *bane* of all trees, while the hemlock and rue *strive against* one another. The olive and the vine "do joy in each others company," and the heliotropic plants have a *love* for the sun.[34]

Regardless of this implicit recognition of some aspects of sentience, the authority of Aristotle's hierarchical tripartite soul led Della Porta to conclude that plants were not fully alive like other creatures. Throughout his descriptions in *Natural Magick*, Della Porta makes repeated reference to two distinct ontological categories, "plants *and* living creatures."[35] This exclusion of plants from the realm of living beings is on the basis of dogma rather than the acceptance of empirical evidence.

Della Porta's influence on Bacon is clear. Comparisons of the *Natural Magick* and *Sylva Sylvarum* reveal that Bacon uses the same anatomical and physiological terminology as Della Porta and similarly aimed his text at the advancement and development of horticultural knowledge.[36] In the botanical sciences, he exhorted the best minds to apply themselves to horticultural and physiological problems in order that the natural world could be more effectively harnessed for the benefit of humans.[37] Bacon had famously strong views about the need to subdue nature for the needs of humankind, "leading to you Nature, with all her children, to bind her to your service and make her your slave."[38] Bacon followed Della Porta in his use of the Platonic/Aristotelian mode of hierarchy and domination in order to reject the fact that plants are living beings. As plants are ecologically dominant in the natural world, I contend that Bacon's beliefs on the nature of plants and their placement in a value-ordered hierarchy are influential on his views of mastery over nature and his exhortation to human kind to engage in unrestricted use of the natural world.

In the sixth book of *Sylva Sylvarum*, Bacon differentiates between the animate and the inanimate, once again using zoocentric criteria. Bacon regarded the presence of "branched veins and secret canals" within which lie the inflamed spirits of living beings, as a way of recognizing animate bodies.[39] For Bacon, plants came somewhere in between the animate and the inanimate. They were clearly differentiated from inanimate things by the ability to feed, grow, and reproduce and through having a clearly defined life span. However, even though they possessed these signifiers of life, the characteristic Western emphasis on difference, and the valuing of reason, led to them being regarded as inferior to "true living creatures." Thus Bacon writes:

> The affinities and differences between plants and living creatures, are these that follow. They have both of them spirits continued and branched, and also inflamed. But first in living creatures the spirits have a cell or a seat, which plants have not, as was also formerly said. And secondly, the spirits of living creatures hold more of flame, than the spirits of plants do and these two are the radical differences.[40]

Bacon identifies two major differences separating plants from "living creatures," and both of them stem from the zoocentric perspective, which is characteristic of the scholastic treatment of plants.[41] Evaluated using an animal framework, plants are considered to *lack* a centralized nature and therefore a place where the animating spirit can reside. As they are of a less active nature than animals, they are assumed to lack the necessary heat required to be properly alive. Instead of relating many of the connectivities between plants and animals, Bacon regards these as "radical differences."[42] He also recounts a list of further differences, which are mainly derived from the works of Aristotle:

> First, plants are all fixed to the Earth; whereas all living creatures are severed, and of themselves. Secondly, living creatures have local motion, plants have none. Thirdly, living creatures nourish from their upper parts by the mouth chiefly; plants nourish from below, namely from the roots. Fourthly, plants have their seed, and seminal parts uppermost, living creatures have them lowermost; and therefore it was said not elegantly alone, but philosophically *Home est Planta inversa*, "man is like a plant turned upwards," for the root in plants is as the head in living creatures. Fifthly living creatures have a more exact figure than plants. Sixthly living creatures have more diversity of organs within the bodies and (as it were) inward figures than plants have. Seventhly, living creatures have sense, which plants have not. Eighthly, living creatures have voluntary motion, which plants have not.[43]

Throughout this list it is clear that all the anatomical features of plants are evaluated from the Platonic/Aristotelian mindset that animals (including humans) are *superior* to plants; a status which for Bacon justified human domination of the natural world. Any differences between animals and plants are listed as *inferiorities* in the case of plants, purely because they differ from the animal mode. This is coupled with an explicit denial of plant sensation and striving. This allows plants to be perceived as an animate kind of *nonlife*, as automatons.

In order to maintain this view of plants as automatons, the clearly purposeful behavior displayed by plants has to be explained away. Bacon was aware of a wide range of plant movements, which could have upset the Aristotelian hierarchy, such as the movements of heliotropic plants or the movement of germinating seeds toward water. Instead of recognizing this sensation and purpose, Bacon describes the movements of plants as being merely mechanical:

> Of this there needeth no such solemn reason to be assigned, as to say that they rejoice at the presence of the sun, and mourn at the absence thereof. For it is nothing else but a little loading of the leaves and swelling them at the bottom with the moisture of the air; whereas the dry doth extend them. And they make it a piece of the wonder that garden claver will hide the stalk when the sun sheweth bright; which is nothing but a full expansion of the leaves. For the bowing and inclining the head, it is found in the great flower of the sun, in marigolds, wart-wort, mallow flowers, and others. The cause is somewhat more obscure than the former; but I take it to be no other, but that the part against which the sun beateth waxeth more faint and flaccid in the stalk, and thereby less able to support the flower.[44]

Even though botany had entered the empirical scientific age, by staunchly maintaining the dogma that plants are passive, Bacon "precluded the concept of sensitivity and retained the hierarchical view of nature."[45] Bacon buried Aristotle's archaic views on plants deep into scientific thought by establishing the position of exclusion as the status quo. In this respect, Bacon's work marks a key threshold in the development of the perception of plants. In other areas of science, Bacon advocated the empirical experimental method for uncovering truth, but in his understanding of plants, he happily and entirely subscribed to the wayward deductive reasoning of Aristotelianism. Bacon failed to call for experimentation on the nature or faculties of plants as the dogmatic subscription to the inferiority of plants fitted well with his designs for enslaving the natural world.

Hierarchy for Mechanists and Atomists

The uncritical acceptance of the Aristotelian hierarchy in the scientific age is not limited to the work of Bacon. A number of influential botanists and philosophers, whose work is often credited with being instrumental in the destructive Western attitude to nature, also implicitly retained the philosophies of Aristotle in their work. The botanist Joachim Jung (1587–1657 CE) is an important figure to consider, for he was instrumental in developing the study of plant morphology, and his work was a major influence on Enlightenment botanists, including John Ray and Linnaeus.

Jung built upon the study of plant morphology by Theophrastus and Cesalpino and developed a comprehensive terminology of morphology, introducing terms to describe leaf margins such as *laciniate*, *crenate*, *serrate*, and *dentate*.[46] Jung was also a committed materialist and atomist, and although his contributions are often overshadowed by those of Gassendi, his work was instrumental in the revival of atomism in the seventeenth century.[47] Jung's work was scientific rather than philosophical in nature, and despite rejecting scholasticism, it is notable that his understanding of the capabilities of plants is based upon the Aristotelian tripartite division of the soul, In his *Isagoge Phytoscopica* Jung writes:

> A plant is a non-sentient body, attached to a particular place or habitat, where it is able to feed, to grow in size, and finally to propagate itself.[48]

Describing this passage, the science historian A. G. Morton notes that "a plant is defined by Jung in words which include, but are not identical with, those of Aristotle—revealing a subtle change of mental attitude and expressing a more physiological view."[49] This physiological view entailed rejecting the *cor* of Cesalpino and dispensing with the terminology of the nutritive plant *soul*.[50] However, for all the changes in terminology, Jung's fundamental understanding of plant existence is still based upon the philosophy of Aristotle. Although he rejected scholasticism, Jung retained Aristotle's basic view that plants are nonsentient organisms. Again the zoocentric connection between a sessile habit and an insentient nature is clear in the work of Jung. While Jung made no proclamations on the design of plants for human use, he propagated the understanding that plants were inferior in nature to animals and human beings.

This dogmatic acceptance of plant passivity and insentience can also be detected in the development of Enlightenment philosophies, which are pinpointed by environmental philosophers as being at the heart of destructive Western attitudes toward nature.[51] Within environmental philosophy, Descartes' dualistic, mechanistic philosophy has been the subject of repeated criticism. Here I do not aim to go over common Cartesian complaints, but instead I simply suggest that the backgrounding of plants may have influenced the development of Descartes' philosophy. In one of the most famous passages of Cartesian thinking, at the end of his *Principles of Philosophy*, Descartes writes:

> I have described the earth, and all the visible world, as if it were simply a machine in which there was nothing to consider but [the] figure and movements [of its parts]. . .[52]

In his *Discourse on the Method,* Descartes posits that the workings of the natural world are underpinned by the laws of Nature, which were established by God.[53] In many instances, Descartes, "fully identified the laws of nature with the laws of mechanics."[54] A brief summary of this physics could be stated as follows. For Descartes, the laws of mechanics explained everything in existence. Indeed Descartes himself even went so far as to state that "my entire physics is nothing but mechanics."[55] As Mathews writes, "The crux of mechanism is its view of the origin of motion," and in Descartes' view, all motion in the universe was applied externally by God.[56] Therefore, in contrast to Aristotle, the general Cartesian position was one that argued for a machine-like Earth, based upon the absence of self-directed, purposeful motion.

So, in this Earthly world, devoid of self-propelled, self-directed movers, Descartes describes the animal body as "a machine made by the hands of God."[57] Significantly this idea famously paved the way for a perception of animals as insentient.[58] Of course, in this respect, human beings were considered an exception. Although in possession of a body that was also considered as a machine, the human differed from the rest of nature by the possession of a distinct mind. Specifically this mind was a consciousness that bestowed sentience and autonomy.[59]

This Cartesian attitude of an insentient, nonvolitional, unminded natural world (standing in contrast to the minded, sentient human) owes something to the Platonic-Aristotelian backgrounding of plants. From his writings, it is evident that Descartes accepted the Aristotelian backgrounding of the plant kingdom. Unlike for animals, Descartes barely deigns to describe plants and does not argue specifically against their sentience; plants are simply lumped together with nonliving entities such as rocks. Although Descartes rejects the general teleology of Aristotle, in this treatment of plants, he retains Aristotle's notion of them as insensitive and nonvolitional.

Garber argues that Cartesian mechanics is most properly seen as an extension of an Aristotelian mechanics, and it is possible that Aristotle's treatment of plants was also influential in the development of Descartes mechanistic view of the machine-like Earth.[60] The uncritical acceptance of Aristotle's botany meant that by default, as the background position, plants were conceived as automatons. Therefore, in combination with the Cartesian view of matter itself as lacking in *telos* and internal motion, the Aristotelian backgrounding of plants (also without volition and motion) would have aided Descartes' argument that self direction was not located in nature, only in human beings. To achieve this, with plants backgrounded, Descartes only had to extend Aristotelian exclusion to the animal kingdom by positing (human) consciousness as the defining aspect of mind. Certainly, if Theophrastean notions of sentient plants had predominated at the time of Descartes, it would have been much harder to conceive of and argue that the natural world was basically inanimate and lacking in *telos*.[61]

This brief treatment of Descartes complements Plumwood's analysis of Cartesian rationalism (and its origins in Platonism) by highlighting that the Cartesian dualism of humans and nature also owes much to the Aristotelian (and Platonic) hierarchical dualisms of animal/plant and human/plant. As well as retaining human beings as the pinnacle of a hierarchy, Descartes accepted the default position of a backgrounded natural world.

The Cartesian idea that nonhumans were basically machines was also aided by the work of Isaac Newton, who published his *Principia* in 1687. The highly technical style of this work rendered it accessible only to a handful of experts, but the effect on the progress of science was enormous. Newton's work introduced the concepts of uniform, mechanical laws for all heavenly and earthly objects, and he is therefore noted as "the man above all who had enthroned the 'mechanical philosophy' in science."[62] Although, Newton's laws are credited with the conception of the natural world as devoid of purpose and direction, it must be remembered that the dogmatic acceptance of Aristotle's doctrine of a passive plant kingdom had rendered much of the natural world as machine like for over a thousand years.

Working within the mechanical framework established by Descartes and Newton, the works of the empiricist John Locke (1632–1704 CE) have been regarded as the "guide and justification for observational and experimental science."[63] Locke's *Essay Concerning Human Understanding* was published in 1690, and championed the philosophy that the only route to understanding and knowledge was gathered by experience and experiment through the reception of sensory information. It is crucial to highlight the fact that, while at the same time helping to establish observation and experimentation as fundamentals in the practice of science, as in the case of Bacon and Jung, Locke dogmatically accepted the scholastic botanical ideas of Aristotle. Thus, Locke helped further entrench an understanding of the plant world that was based upon faulty observations, was zoocentric in nature, and was contrived to exclude plants from human consideration. Regardless of the break from scholasticism, the Lockean views on plants were based upon a drive to exclusion and separation rather than upon reasoned observation or experimentation.

Chapter 27 of Locke's *Essay* owes much to the Aristotelian tripartite division of the natural world in dividing living beings into three distinct ontological categories, emphasizing the separation of the "Identity of Vegetables," the "Identity of Animals," and the "Identity of Man." The Lockean treatment of plants is very brief, but in the *Essay*, Locke distinguishes plants in the familiar Aristotelian manner as living beings that are only "fit to receive and distribute nourishment."[64] Again, rather than on the basis of experiment or observation that Locke so championed, it is clear that plants are entrenched in a position of exclusion on the basis of faulty Aristotelian reasoning and an extended emphasis on human superiority and separation.

Following Aristotle and Descartes, Locke only attributes any facet of rationality to human beings. This rationality was dependent upon self consciousness. In the *Essay*, Locke decrees this self consciousness necessary for a sense of personal identity, "for, it being the same consciousness that makes a man himself to himself."[65] On the basis of the plant kingdom being denied autonomy and individual identity, Locke treated the wider natural world in the same fashion. In the *Second Treatise of Government* Locke describes America as a "vacant" place, a place that is akin to the original natural world, given by God for man to use and to exercise his labor upon (see Chapter 3).[66]

Therefore, as for Descartes, the mechanism of Locke did not radically alter the perception of plants that had been in existence since Aristotle. The Cartesian and Lockean dismissal of the possibility of mind and striving in the natural world are strong exclusions which continue to resonate, but they can be considered as reiterations of the exclusions and hierarchies found in the works of Plato and Aristotle. By studying the pervasive spread of Aristotle's influence we can detect that a priori assumptions of plants being without inductive sensitivity and intelligence are actually embedded in the scientific philosophy of mechanism. The introduction of this mechanistic philosophy by Bacon, Descartes, Newton, Locke et al. is significant for entrenching and legitimating the idea that plants are passive and naturally inferior, thus cementing and extending the norm of exclusion.

Systematics and Hierarchies

Despite the steady accrual of botanical evidence to the contrary (see Chapter 7), this zoocentric/anthropocentric mode of viewing plant life was also sanctioned by the most important botanists of the Enlightenment period. As for more philosophical thinkers, one of the fundamental reasons for the failure of prominent botanists to counteract the notion of plants as passive resources is the dogmatic strength of a hierarchy in nature, also found in the Bible (see Chapter 3).

John Ray (1627–1705 CE) is regarded as the most influential botanist of a crucial period in the late seventeenth century. Ray's work greatly advanced systematics as well as plant anatomy and physiology. Morton considers Ray the "founder of plant physiology."[67] In the work of Ray, it is possible to detect the philosophical ideas of mechanism colliding with ancient scholasticism. Ray was strongly influenced by Theophrastus, to the extent that he retained the Theophrastean taxonomic division of trees, shrubs, sub shrubs, and herbs, despite the fact that this was dismissed as artificial by Jung.

Just as John Ray's *Historia Plantarum* is important for plant taxonomy, so his *Wisdom of God* is vital for an understanding of the continued influence of the Aristotelian position. From the outset, *Wisdom of God* elaborates Ray's views on the status of plants. When discussing the relative abundance of plants and animals in the world, he remarks that nature is "more sparing in her more excellent productions."[68] It is clear that, following Plato and Aristotle, Ray considers plants to be inferior to animals and human beings. The direct influence of Aristotle is starkly illustrated in the following passage:

> Animate bodies are either endued with a vegetable soul, as plants; or a sensitive soul, as the bodies of animals, birds, beast, fishes and insects; or a rational soul, as the body of man.[69]

Nearly two thousand years after it was first elaborated, the authority of Aristotle was still ensuring the dogmatic repetition of the tripartite division of the natural world. It is apparent that Ray accepted without question the "fact" that plants were passive and insensitive, and so were subservient beings.

As for Bacon and for Locke, this passivity was a natural indication of the subservience of plants. *Wisdom of God*, as the title suggests, was written in praise of God the "intelligent creator" and shaper of the natural world.[70] Ray argues for the teleology of design throughout this work and argues that plants and animals must have been created by God for human beings, otherwise "he doth but abuse them, to serve ends for which they were never intended."[71] It is clear from Ray's argument that the stripping away of plant autonomy is crucial for maintaining domination of the plant kingdom.

Again, as did Jung, Bacon, Locke, and Descartes before him, Ray made it clear that he rejected Aristotelian terminology, while he failed to reject the attitude of exclusion and separation in Aristotle's descriptions of the plant kingdom.[72] Ray's descriptions make it clear that passivity was commonly accepted, yet at the same time his own observations also recognize an autonomous, active, knowledgeable, organizing principle in plants:

> For what account can be given of the determination of the growth and magnitude of plants from mechanical principles, of matter moved without the presidency and guidance of some superior agent? [73]

> That all this can be done, and all these parts duly proportioned one to another, there seems to be necessary some *intelligent plastic nature*, which may understand and regulate the whole economy of the plant, for this cannot be the vegetative soul, because that is material and divisible together with the body.[74]

Influenced by the Cambridge Neoplatonists, Ray uses the term *intelligent plastic nature* to try and explain active behavior.[75] Yet in the very same work, he explains the heliotropic movements in purely mechanical terms. Ray regards movement to be achieved by wilting of the stem after heating by the sun, thus retaining the perception of plants as automatons. In order to retain the notion of plants as passive, he also uses mechanistic terminology to explain the movements of the remarkable touch-sensitive plant *Mimosa pudica*:

> For when cooled or squeezed by the finger, the expanding motion of the sap in the nerves is prevented; whence suddenly the fibres shorten themselves and the pinnae of the leaves are drawn together, just as the same cold compresses and folds up the skin of our bodies, as on the contrary, heat generally extends it.[76]

It is important to note that the entrenched position of passivity necessitates this mechanistic explanation, as plants had a priori been denied the capability of sensation and movement. Therefore even though plants are described as possessing an intelligent plastic nature, they are not recognized as intelligent and purposeful.

Ray's attacks on the prevailing Cartesian ideas of the time demonstrate the extent of plant backgrounding. When Ray beseeches his readers not to be taken in by Descartes' argument that only humans have sense perception and reason, his counter arguments concentrate on a defense of the existence of sensation and intelligence in animals. Ray does not attack Descartes (implicit) view that plants are insentient, as this idea was merely the expression of an entrenched position of exclusion that most took for granted. Unlike animals, plants had already been rendered as mere vegetables for a very long time.

As for those Enlightenment thinkers before him, Ray's maintenance of this exclusion is connected to the subjugation of the plant kingdom for human use. Ray was greatly influenced by Pliny, and agreed with Pliny, Plato, and Aristotle that plants were "designed for the food of animals."[77] As Chapter 3 argues, this teleology was strengthened by similar hierarchies that appear in the book of Genesis. Subscription to this theory of creation necessitates the idea that plants are naturally subservient and inferior to animals. Ray expresses this inferiority by asserting that unlike for animals, "no man is troubled to see a plant torn, or cut, or stamped or mangled how you please."[78] Nowhere does he question why this is the case.

In addition to Ray, the dogmatic acceptance of plant inferiority can also be seen in the works of Linnaeus (1707–1778 CE). Linnaeus had little interest in plant physiology, yet in *Systema Naturae*, he divides the natural world into plant, animal, and mineral and explicitly ranks these ontological categories in terms of value and ability. The influence of Platonic and Aristotelian hierarchies is obvious in Linnaeus's declaration that "the animal kingdom ranks highest in comparative estimation, next the vegetable, and the last and lowest is the mineral kingdom."[79] For Linnaeus, plants are inferior as they are not capable of sensation or reason:

Vegetables clothe the surface with verdure, imbibe nourishment through bibulous roots, breathe by quivering leaves, celebrate their nuptials in a genial metamorphosis, and continue their kind by the dispersion of seed within prescribed limits. They are bodies organised, and have life and not sensation.[80]

As with all previous attempts, a feature of the Linnaean hierarchy is an emphasis on the differences between plants and animals. Plants are evaluated from a zoological perspective, considered only to be *lacking* attributes possessed by animals. In Linnaean systematics, it is not subjective lives that are under scrutiny, but simply objects to be filed away in a system that would enable their easier use by human beings.[81] As is reiterated throughout this chapter, the effect of this process of backgrounding and exclusion is justified domination of the plant kingdom and the wider natural world. The connection between exclusion and domination is clearly expressed in the work of Comte de Buffon (1707–1788), who although a fierce critic of Linnaeus, did not dispute that:

[Man] is master of the vegetable tribes, which, by his industry he can, at pleasure, augment or diminish, multiply or destroy. He reigns over the animal creation; because like them, he is not only endowed with sentiment and the power of motion, but because he thinks.[82]

Although the dogma of passivity and subservience is contradicted by careful observation of plant behavior, I contend that it has been maintained partly in order to justify untrammelled resource use by human beings. Stripping away autonomy and sentience allows plants to be seen "as mere bodies, and thus as servants, slaves, tools, or instruments for human needs and projects."[83] The backgrounded position of plants has also remained mainly unquestioned by easily erupting into dogma when threatened. The following chapter puts forward the case that the exclusion and the domination of the plant kingdom remained unquestioned because it is also authorized by the West's dominant religious tradition.

3

Passive Plants in Christian Traditions

> And of every living thing of all flesh, you shall bring two of every sort into the ark to keep them alive with you. They shall be male and female.[1]
>
> —Genesis 6

In his seminal paper *The Historical Roots of our Ecological Crisis,* Lynn White Jr posited Christianity as the "root cause" of our current environmental crisis and argued that Christianity is the "most anthropocentric religion the world has seen."[2] For White, the fact that the Bible sanctions dominion over the natural world is reason for Christianity to have a "huge burden of guilt" with regard to the current environmental predicament. Interpreting biblical passages from Genesis, White's thesis posits that the Bible sanctions man's position as a despot, ruling over creation as he wishes.[3] In the words of James Barr, "as God is the lord over His whole creation, so He elects man as His representative to exercise his lordship in God's name over the lower creation."[4]

Unsurprisingly, White's thesis has catalyzed the emergence of a Christian ecotheology, which has aimed to address these charges.[5] Most significantly, the dangerous word *dominion*, which is used in Genesis 1, has been recast by Christian ecotheologians as "stewardship," a concept which is claimed to acknowledge human difference, without sanctioning unrestrained use of nature.[6] In stewardship, it is claimed that although human beings are set apart from nature in one respect, by being granted control of and responsibility for other created beings, the Bible also shows a clear "horizontal relationship in which humans relate to their fellow creatures as all creatures of the one Creator."[7] Theologian Jürgen Moltmann expresses this relationship as a type of solidarity of creation, with humankind standing "together with all other earthly and heavenly beings in the same hymn of praise of God's glory."[8] A similar interpretation is shared by H. Richard Niebuhr, who argues that because God is the origin of creation, God is the "center of value," and therefore, all things have value in their relationship with God. Thus, as equal products of a creative God "Sparrows and sheep and lilies belong within the network of moral relations. . . ."[9]

This position is also taken by Callicott in a review of the diversity of biblical interpretations: as created beings are said to be "good," this is set down as evidence that these beings possess independent value and intrinsic worth.[10] On this basis, Callicott has advocated the stewardship interpretation of human-nature relationships.[11] In support of this, Tom Hayden claims that the fact that "it is based on scant empirical evidence and a debatable reading of scripture may be irrelevant."[12] Yet, while stewardship (and its notions of responsibility and accountability) is certainly preferable to domination as a mode of relating to the natural world, the evidence for a nondominant account of the natural world in the Bible is indeed scanty. A study focussed on plants also reveals that it is limited to discussions of the animal kingdom. Most often, domination, subjugation, and anthropocentricity in Christianity have been refuted using biblical examples that demonstrate a concern for animal welfare.

For scholar Robin Attfield, passages in the Old Testament that "a right minded person cares for his beast"—and in the New Testament, Jesus's general concern for animals—suggest that the Bible is "irreconcilable both with an anthropocentric ontology and with anthropocentric accounts of value."[13] Similarly, looking at the Old Testament from within the Jewish tradition, Eric Katz regards the fact that Genesis 1:29 restricts the human to the use of plants as a clear rebuff to the notion of "dominion, control and ownership of all the living creatures in nature!"[14] Similarly, Moltmann rejects the charge that Genesis sanctions domination of the natural world based upon the relationship between human beings and animals.[15] Discussing the relevant passage in Genesis 1:26, he dismisses the notion of an "anthropocentric dominion." This is based on the fact that human beings are originally only allowed the plants of the Earth and are not sanctioned to eat other animals. Therefore, for Moltmann:

> The rule of human beings over the animals can only be a rule of peace, without any "power over life and death." The role which human beings are meant to play is the role of a "justice of the peace."[16]

However, a detailed study of the nature of Christianity's attitude to, and relationship with, the plant kingdom is conspicuously absent from discussions of the wider attitude to nature. Even in Habel's Earth Bible series, which aims at restorative ecojustice and ascertaining "whether Earth and the Earth community are oppressed, silenced, or liberated in the biblical text," there is no in-depth portrayal of plant life.[17] The omission of plants from the Earth Bible and the debate on Christian attitudes to nature is significant because, as discussed previously, plants make up the bulk of visible nature and form the basis of our human reliance upon the natural world. It is significant because if the Bible sanctions domination of the plant kingdom, it sanctions domination of much of the natural world and those creatures that directly sustain human life.

Before restorative ecojustice can be achieved for the plant kingdom in the Christian tradition, it is important to uncover the dominant human-plant relationship and show where plants have been oppressed and silenced. Detailing human-plant relationships within Christian sources is a significant project in itself, because while biblical passages may attribute some consideration to animals and indicate possible "horizontal relationships" between humans and animals, there is little evidence that such consideration is given to plants. Indeed, evidence that can support the notion of horizontal, or heterarchical, relationships between plants and human beings is extremely scarce. A careful reading of biblical and theological material makes it clear that the very opposite is the case. As in Platonic and Aristotelian philosophies, plants are subjected to the processes of backgrounding and domination.[17] In a range of Christian sources, plants are constructed as the radical Other, the underside of an ontological dualism that is essentially zoocentric in nature.[18]

The Genesis of a Radical Divide

The stories of creation and destruction in the book of Genesis are integral to the portrayal of plants in the Bible.[19] In the first biblical book, we encounter the Hebrew/Christian God, both an ethereal and materially creative being, whose first act is to manifest the heavens and the Earth. Genesis 1 describes God as the creator of light and water on the first two days, and the creator of plants and animals later on in the week. In this respect, all plants, animals, and human beings share God as their origin and this must be recognized as a possible basis for the existence of "horizontal" or heterarchical relationships between plants and human beings.

Further evidence for such relationships has been garnered by theologians with panentheistic tendencies. This has been used in order to demonstrate that not only is God the origin of all created things, but that part of God also exists in all his created works. In the Old Testament and the Hebrew Tanakh, the creative power of God stems from the spirit or breath (*ruach*) of God sweeping over the Earth. In his attempt to manifest a "return to the original truth" of the Biblical teachings, Moltmann interprets this causal existence of God's spirit as evidence for the immanence of God in the natural world.[20] For Moltmann, "from the continual inflow of the divine Spirit *ruach*, created things are formed," and there is a further assumption "that this spirit is poured out on everything that exists, and that the Spirit preserves it, makes it live and renews it."[21] Therefore, in the panentheistic movement, "through his cosmic Spirit, God the Creator of heaven and earth is present *in* each of his creatures and *in* the fellowship of creation which they share."[22]

However, while Moltmann's interpretation focuses on the fellowship of created beings, it fails to address the description of sharp ontological differences described in the accounts of plant, human, and animal creation. Plants are described in Genesis 1 as being created by God on the third day of his activity. The animals were brought into being on the fifth day of creation. In these oft quoted passages from Genesis, it is stated:

> And God said, "Let the earth sprout vegetation, plants yielding seed, and fruit trees bearing fruit in which is their seed, each according to its kind, on the earth." And it was so. The earth brought forth vegetation, plants yielding seed according to their own kinds, and trees bearing fruit in which is their seed, each according to its kind. And God saw that it was good. And there was evening and there was morning, the third day.[23]

> And God said, "Let the waters swarm with swarms of living creatures, and let birds fly above the earth across the expanse of the heavens." So God created the great sea creatures and every living creature that moves. . . .[24]

> And God said, "Let the earth bring forth living creatures according to their kinds—livestock and creeping things and beasts of the earth according to their kinds." And it was so.[25]

These descriptions of the origins of plant and animal life in Genesis 1 indicate that both animals and plants share some sort of kinship because they are both derived from the Earth. As Hillel points out, this kinship is even extended to humankind in the naming of the first human from *adamah* (earth).[26] However, despite the subtle connectivities in this notion of a shared origin, they are significantly challenged by an ontological split between plants, animals, and humans.

In particular, for a treatment of plants, it is of great significance that the excerpt concerning the origin of zoological life, the animals of sea and earth are described as "living creatures." However, in the passage that deals with the origin of plants, there is no description of plants as living beings. In fact, in this opening account of creation, I would claim that plants are backgrounded as inanimate, nonliving beings.

This ontological schism between plants and animals is both reiterated and deepened in the well-known story of Noah and his ark. The story of Noah opens with God deciding to destroy all living beings on the Earth as a punishment for human wickedness. To accomplish this destruction, he calls forth a great flood:

> For behold, I will bring a flood of waters upon the earth to destroy all flesh in which is the *breath of life* under heaven. Everything that is on the earth shall die. But I will establish my covenant with you, and you shall come into the ark, you, your sons, your wife, and your sons' wives with you. And of *every living thing of all flesh*, you shall bring two of every sort into the ark to keep them alive with you. They shall be male and female.[27]

This passage simultaneously confirms, deepens, and justifies the biblical notion of plants as nonliving beings. In this passage, the fundamental definition of a living being is one that has both *the breath of life* and *flesh*. As Moltmann recognizes, this distinction is made on the basis of the Hebrew terms for the vital principle or soul (*nepes* or *nephesh*).[28] The Hebrew root of these words for living or ensouled beings means "to breathe," and so based upon this conception of life within the biblical text, breathing is regarded as "a decisive mark of the living creature."[29] As the term *nepes* "denotes the total person," it is also linked to the circumscription of personhood, and in this context, *nepes* is used in the Bible as a synonym for the personal and reflective pronouns.[30]

Such an interpretation is borne out in the details of the great flood story. God instructs Noah to collect specimens of the living creatures, which will be able to repopulate the Earth once the flood has receded. In the context of this rescue job, however, plants are not designated as living beings. Therefore, not a single individual from the world's three hundred thousand flowering plant species is taken onto the ark.[31]

It is possible that in one sense this is an expression of the greater vulnerability of animal life to the floodwaters.[32] However, it must be remembered that in an agriculturally sophisticated culture such as ancient Israel, people would have been well aware of the vulnerability of many plants to floodwaters, both freshwater and saltwater.[33] Instead, as plants are portrayed as beings without life, the floodwaters are of no concern to the plant kingdom. It is claimed that the flood "blotted out every living thing that was on the face of the ground."[34] As all plants apparently survived the complete (150 day) submersion in saltwater, the text makes it abundantly clear that plants are not capable of being harmed and are not considered to be living. Indeed plants are only included in the rescue of living beings when God informs Noah to construct the ark out of *gopher* wood.[35] This floating wooden ark made from the parts of nonliving plants and filled with living animals encapsulates the radical ontological divide between plants and animals. A more appropriate ecological escape plan would have included plants on the ark.

The hypothesis of a radical division between plants and animals is supported by material from the story of creation in Genesis 2–3. This second account of creation is several centuries older than that of the Jewish priests and differs from it considerably.³⁶

> The lord God formed the man of dust from the ground and breathed into his nostrils the breath of life, and the man became a living creature. And the lord God planted a garden in Eden, in the east, and there he put the man whom he had formed. And out of the ground the lord God made to spring up every tree that is pleasant to the sight and good for food.³⁷

In this account of the creation, human beings are given the breath of life (*ruach*) from God, and it is this breath that makes man a living creature (*nepes*). Plants however, are not treated so favorably by God and are denied the element which would render them living. Animals are described as living creatures, but are not directly formed by God, or like man, directly given God's breath. Thus, man is made of the same substance, but is created and animated differently.³⁸ Although Moltmann rightly highlights that plants are formed from the creative spirit or breath of God (*ruach*), this passage makes clear that plants are not invested with the more personal, animating, connecting aspect of this breath.³⁹

As in Plato and Aristotle, this portrayal of plant life in Genesis casts plants as radically different from animals and humans. They are cast as the Other on the basis of *lacking* certain attributes such as *breath* and *flesh*. However, this biblical rendering of plants goes much further in its backgrounding. It totally denies life, rather than just sensation and mind.

The above passages relate that to be considered as *living*, beings should move and breathe. They should also have flesh and blood. In Genesis 9, this link between blood and life is made explicit when thanking Noah for rescuing Earth's animals, God allows Noah and his descendants to eat animal flesh for the first time:

> Every moving thing that lives shall be food for you. And as I gave you the green plants, I give you everything. But you shall not eat flesh with its life, that is, its blood. ⁴⁰

Once again, the existence of life is directly related to the possession of breath, blood, and flesh and indirectly with movement. As plants are sessile—do not possess obvious breathing and do not have flesh and blood in the same way as animals—they are rendered as inanimate.

This approach stands in direct contrast to other cultures that have recognized the life and indeed mentality of plants on the basis of qualities and attributes such as birth, growth, illness, decay, and death.⁴¹ The occurrence of these processes in plants can scarcely have escaped the notice of ancient Hebrews. The growth and decay of plants must regularly occupy the thoughts of any agrarian society. It is clear that this evaluation of life as one lived with breath and flesh only occurs as a result of having chosen criteria that intentionally exclude plants. Rather than demonstrating an etiological explanation of the way things are, it is my interpretation that this is a political construction of the way some humans want the natural world to appear. This backgrounding position could have been arrived at through slow, gradual slippage, but it likely became the default position through convenience.

As with Platonic and Aristotelian philosophies, defining animals (and humans) in opposition to plants appears as a deliberate attempt to remove the connectivities between animals, plants, and humans. Again this can be clearly located in Genesis 1, when God decrees plants as being in existence solely as food for animals and human beings:

> And God blessed them. And God said to them, "Be fruitful and multiply and fill the earth and subdue it and have dominion over the fish of the sea and over the birds of the heavens and over every living thing that moves on the earth." And God said, "Behold, I have given you every plant yielding seed that is on the face of all the earth, and every tree with seed in its fruit. You shall have them for food. And to every beast of the earth and to every bird of the heavens and to everything that creeps on the earth, everything that has the breath of life, I have given every green plant for food."⁴²

In this famous biblical passage, the radical ontological schism between animals and plants is reiterated. Animals (including humans) that have the breath of life, and other signifiers of life such as blood, are rendered as naturally superior to plants. In contrast with the interpretation of Moltmann, it is clear that this situation leaves animals (especially humans) as the rightly dominant force over much of the natural world. Therefore instead of also recognizing autonomy and self-directed purpose in the plant kingdom, Ezekiel 47:12 only says of plants that "their fruit will be for food, and their leaves for healing."

As in the philosophies of Plato and Aristotle, it is important to consider that this rendering of plants as bereft of the faculties of animals and humans—in this case bereft of life—is strongly connected to the insistence on a natural hierarchy of value in the world. For Anne Primavesi, it is the drive toward hierarchical value-ordering that underpins the "Christian binary codes of heavenly/earthly; sacred/secular; human/animal; male/female."⁴³ To this list, we may wish to add the dualisms of animal/plant and, especially in a human-penned document, human/plant.

Within Genesis, there is a subtle hierarchy of the natural world based upon the relative similarities of created beings to the creator. Within the natural world, humans and plants occupy different ends of a hierarchy. This hierarchy is rejected by Moltmann who recognizes a special human place in creation but denies that humans are at the apex of creation. In this place, he positions the Sabbath as the pinnacle of God's creations.[44] However, in the context of real world relationships between beings, it is a fact that humans more closely resemble God than any other creature. Only human beings are made in God's image.

Plants are deemed to be much further away and more unlike God. On this premise, plants are placed at the bottom of a remarkably anthropocentric representation of natural value.[45] Further from God, according to the analysis of Primavesi, plants are then implicitly regarded as being of far lesser value than other created beings. In connection to this, an important insight is that once this hierarchy has been established, God functions in relation to those at the bottom of the hierarchy as "validation of their subordination and consequent exploitation."[46] Again, the existence of this rigid, suppressive hierarchy gives rise to a political biblical interpretation.

Symbolic Similarity

Countering this argument, there are a number of passages involving plants in which they *are* equated with living beings and which suggest that they possess agency, autonomy, and volition. It is possible that the backgrounding of plants is counterbalanced by passages that appear to focus more on similarity than on difference. Indeed plants appear to be strongly linked with life and vitality in several biblical passages that directly associate plants and people.

A common association appears in the use of the plant to portray humanity. In Amos, the body of the Amorite is described using a plant metaphor; the scripture reads, "I destroyed his fruit above and his roots beneath."[47] The Song of Solomon is similarly poetic when the bride describes her husband:

> As an apple tree among the trees of the forest, so is my beloved among the young men. With great delight I sat in his shadow, and his fruit was sweet to my taste.[48]

Isaiah also uses a similar human-plant simile to describe the transience of human life. He proclaims that "all men are like grass, and all their glory is like the flowers of the field."[49] As well as transience, the small herbaceous plants are also used as images of vitality. Eliphaz counsels Job that "you shall see your seed multiply, your offspring like herbs on the earth."[50] The Psalms also use the image of a vigorous herb to portray both the vitality and transience of a human being, "he sprouts like a herb and flourishes only later to be destroyed."[51] This message of transience in the plant kingdom is echoed by Job who points out that man "blossoms like a flower and then he withers."[52] Job also takes this link between plants and humans even further by describing himself as a tree.[53] This sentiment is echoed by Jeremiah who regards the righteous man as a planted tree, which stretches his roots to the stream and does not fear heat.[54]

Not only are plants portrayed as alive; they also appear as aware, communicative, perceptive, and autonomous beings. The Parable of the Trees in Judges 9 is a particularly interesting passage because it imparts a subjective nature to trees and highlights their faculties of communication and choice. In this passage, the trees of the land search for a king and decide to select one from among them to rule. The parable details how the trees discuss their choices and attempt to persuade a succession of trees to take over as king, until finally settling on the tree *Ziziphus spina-christi* (L.) Desf. The Bible also provides some more poetic interpretations of the lives of plants:

> The wilderness and the dry land shall be glad; the desert shall rejoice and blossom like the crocus; it shall blossom abundantly and rejoice with joy and singing.[55]

This notion that plants can express their appreciation to God also occurs in the Psalms.[56] Zachariah appears to portray the expressive nature of trees:

> Wail, O cypress, for the cedar has fallen, for the glorious trees are ruined! Wail, oaks of Bashan, for the thick forest has been felled! [57]

Yet, for all the association between plants and living, autonomous beings, it is clear that the attribution of plants with these qualities remains entirely allegorical in the sense of meanings carried across an ontological divide. It is the fact that in the general order of things, trees do not wail, that gives this poetry its power. Where plants are ascribed qualities such as perception and sensation, it is clear that these metaphors firmly locate such attributes first and foremost in the human being. Many tree metaphors, such as those found in Isaiah are not intended to elucidate the reader on the relationship between humans and the natural world, but are instead aimed at imparting theological views to the reader.[58] Plant metaphors occur frequently in the Bible, but in the association of plants with living beings, plants are used as *symbols* of life and are not actually and explicitly recognized as being alive.

Instruments of Agriculture

Although Hillel has highlighted the environmental and textual heterogeneity of the document we know as the Bible, across the range of Biblical sources and ecosystems, the human-plant relationships are consistently characterized as instrumental.[59] This preference for utilitarian relationships is evident in the creation story of Genesis 2–3. In Eden, the plants that God places in the garden are all useful to human beings in some way. Throughout the Old Testament, the focus is on domesticated plants for food and for shelter. Plants are not beckoned to form a wild or natural setting, inhabited by non-useful, straggly, or messy vegetation. Only useful plants are found in Eden, while other plants are left outside the boundaries.

The same principle also applies in other parts of the Bible. Although Middle-Eastern botanist Michael Zohary depicts the Bible as "the most pervaded with nature of all scriptures or ritual historical works,"[60] plants are overwhelmingly portrayed as instruments for human use. The vast majority of plant references occur in a horticultural or agricultural context. Initially this may not seem unusual. The lands of the Middle East in which the Israelties lived were those in which agriculture first began to flourish. The point I wish to make is not a criticism of agriculture, but simply the fact that with the autonomy of plants vitiated, there is no counterbalance to an agricultural preference for *domesticated* or *subjugated* plants. There is no celebration of free-living autonomous plants. Jeremiah's recognition of the preference for well-maintained crops[61] may be shared by more animist peoples such as the O'odham people of southern Arizona, however there is no counterpoint to this preference for cultivars.[62] Where the O'odham recognize that wildness is linked to health, wholeness, and liveliness, there is no such recognition in the biblical treatment of the plant kingdom.[63]

In modern day Israel/Palestine there are approximately 2,700 plant species, yet only 110 plant species actually appear in the Bible. Many of these are described only once and the vast majority of the 110 species that are mentioned are agricultural plants.[64] Human survival depends upon the consumption and use of plants, and the life that agricultural plants bring should rightly be celebrated. Indeed, many of the situations and ecosystems in which the Israelites found themselves were tough, requiring a great focus on effective agriculture.[65] The crucial point, however, is this; such an emphasis on agriculture is based upon the divine decree that plants are without life, autonomy, and agency. It is based upon the established hierarchy of value set down in the Genesis text. Therefore, it is a description of an agricultural setup in which plant life has no other purpose. The only thing of importance is the satisfaction of human needs. Such a perspective can be found in a description of the promised land from the Old Testament.

In this definitive description of the land of Canaan, hundreds of plant species are represented by a list of just seven agricultural species:

> For the lord your God is bringing you into a good land, a land of brooks of water, of fountains and springs, flowing out in the valleys and hills, *a land of wheat and barley, of vines and fig trees and pomegranates, a land of olive trees and honey*. . . .[66]

These seven plant species are the most important plants from a human agricultural perspective, and understandably are cherished in this passage. However, instead of more dominant natural plant communities such as oak or cedar woodland, in this passage, it is noticeable that these crops actually *define* the land and the flora.

Bible passages demonstrate that it is the agricultural crops which are the most revered in life by both God and by humans. The agricultural plants are directed by God to produce high yields for the sake of humans, commanding the land to "put forth your branches and yield your fruit for my people Israel."[67] The orientation of humans in the Bible is solely toward the maximization of plants as resources rather than combining the need to eat with a respect for the purposes and needs of other beings. Human activity is directed solely toward an abundance of crop plants:

> Behold, the days are coming, declares the lord, when the plowman shall overtake the reaper and the treader of grapes him who sows the seed; the mountains shall drip sweet wine, and all the hills shall flow with it.[68]

Those plants in the natural world that are not useful for humans are placed firmly at the bottom of a hierarchy of value and importance. This is clearly seen in the Old Testament in a section of the law code in the book of Deuteronomy, specifically the chapter which sets out the rules of war for the Israelites. This passage advises:

> When you lay siege to a city for a long time, fighting against it to capture it, do not destroy its trees by putting an axe to them, because you can eat their fruit. Do not cut them down. Are the trees of the field people, that you should besiege them? However, you may cut down trees that you know are not fruit trees and use them to build siege works until the city at war with you falls.[69]

The trees that are useful for agricultural purposes are spared the axe, not in any way because they have life and autonomy but entirely because they can be used to sustain the men of the attacking army. Trees whose uses aren't so strategically apparent are not deemed worthy of any sort of protection and can be felled with impunity. This treatment of plants is also seen in the New Testament. A good example is the Parable of the Barren Fig Tree in which Jesus describes the actions of a man who planted a fig tree in his vineyard. At the end of the summer, the man found that there was no fruit on his fig tree and so he said to one of his workers:

> Look, for three years now I have come seeking fruit on this fig tree, and I find none. Cut it down. Why should it use up the ground? [70]

In this passage, it is clear that the man cannot see any reason why the tree should exist without providing for human wants. Further on in the text, the man's servant attempts a defense of the tree, but even this is articulated on the basis that the tree will surely have instrumental value in the future.

Such an instrumental view of plants even extends into the actions of Jesus in the New Testament. In a passage from Mark, Jesus is hungry during the Passover, and although early spring is obviously not the season for figs, Jesus approaches a fig tree seeking fruit. When he finds only leaves covering the tree, he curses the fig tree and issues the curse, "May no one ever eat fruit from you again."[71] When his disciples pass by the following day, they find that the cursed tree has withered and died. Without its fruits, even to Jesus, the tree has little worth. Bereft of life and autonomy, its value lays only in what it can provide to human beings. Entangled in these instrumental relationships with plants, Jesus was not acting in any way immorally when he made it wither. In fact he was following divine precedence as God himself had a habit of attacking the plants of Israel with disease in order to gain the obedience of the people.[72]

Rendered as inanimate, plants are deemed unsuitable for care-based, dialogical relationships with human beings. To illustrate this, the only place in the Bible where a plant is the recipient of compassionate feeling acts as an example of inappropriate behavior toward the plant kingdom. The book of Jonah relates that on a hot day, Jonah was sat outside under the blaring sun and in order to provide shade, God sent him a plant, possibly a castor oil plant, *Ricinus communis* L. Jonah was happy about the respite from the sun but when on the next day God sent a worm to eat the plant and Jonah lost his shade and became angry. Interestingly, Jonah is not just angry about the loss of his shade, but also about the harm done to the plant itself. This compassion for plant life is unique in the Bible, yet it is clear that this care for the plant is not sanctioned. God rebukes Jonah for pitying the death of the plant and instead directs him to feel compassion for the humans and animals of the city.[73]

Christian Theology and St. Augustine

Exploring the work of perhaps the two most influential thinkers in Christian theology uncovers a continued drive toward constructing plants as radically inferior to humans. It demonstrates the relative importance of the natural hierarchies found in the stories of creation versus the metaphorical linkages between plants and human beings. The focus on the hierarchy of creation in Christian theology is undoubtedly amplified and consolidated by the pervasive influence of the authoritative philosophies of Plato and Aristotle.

The combined expression of biblical and Greek hierarchies in the natural world is prominent in the work of Christian theologian St. Augustine (354–430 CE). Augustine's theological work is of great historical importance, and although it is some of the earliest in the history of the Church, it remained fixed for over one thousand years until the Reformation, when it was favored by Protestants and rejected by Catholics.[74] Although new scholarship demonstrates relational aspects to his theology, Augustine is important because in his writings plants are explicitly excluded from ethical relationships with human beings.[75]

In his most important works, Augustine attempts to reconcile ancient Greek thought with Christian scripture. Augustine's understanding of the soul appears to be a mixture of biblical, Platonic, and Aristotelian influences.[76] An excerpt from the *City of God* mirrors Aristotle's version of the faculties of the soul, and clearly recognizes that plants are living beings:

> He made man a rational animal composed of soul and body. . . . He gave life capable of reproducing itself in common with the trees; and sentient life, in common with the beasts; and intellectual life, in common with the angels alone.[77]

In breaking with the biblical assertion that plants are inanimate, Augustine follows Aristotle's tripartite division of the faculties of the soul. In this theory of the natural world, plants are considered to be capable only of reproduction and growth; sentience is denied to plants but given to animals. Man is the only earthly being considered to have the faculty of intellect. Yet, after delineating these faculties based upon Aristotle, Augustine then follows Plato by speaking of two types of soul, the irrational and the rational[78]:

> To the irrational soul He has given memory, sensation and appetite; and to the rational soul, he has given mind, intelligence and will.[79]

From the earlier passage it is clear that when talking of a rational soul, it is only human beings which possess such a soul. Only human beings are deemed capable of intelligent behavior.[80] This drive toward separation and placing human beings as the pinnacle of the natural world is characteristic of both the biblical and Greek philosophical descriptions of creation. So too is the backgrounding and subjugation of plants which St. Augustine accomplishes by denying plants even the possession of the irrational soul.

Even though he was influenced by Aristotle's tripartite theory of the soul, Augustine asserts that plants are without soul on the strength of the creation stories in Genesis. Augustine clearly links the possession of breath not with the existence of life, but with the possession of soul. Using Genesis 1–3, he relates that when shaping man and animals, God endowed them with breath, and "God's breath, which He made by breathing . . . is itself the soul."[81] Although plants are recognized as being alive, as they are not given God's breath, they are not ensouled. In equating breath with soul, Augustine manages to retain the radical difference and inferiority of plants.[82]

As with Aristotle before him, Augustine achieves this depiction of plant life by glossing over points of connection in his schemata of living beings. In two important passages in the *City of God*, Augustine recognizes that plants "seek in their own fashion to conserve their existence, by rooting themselves more and more deeply in the earth" and that a plant's "nourishment and generation have some resemblance to sensible life."[83] However, these connectivities are not developed as the basis of relationships between living beings. As in biblical and Aristotelian thought, the hierarchical value-ordering of the natural world predominates.

The connection between backgrounding, inferiorization, and the drive to dominate the natural world as a resource, appears in a passage from the *City of God*. In an attempt to refute the commandment "thou shalt not kill" being extended to animals and plants, Augustine argues:

> When we say, Thou shalt not kill, we do not understand this of the plants, since they have no sensation, nor of the irrational animals that fly, swim, walk, or creep, since they are *dissociated* from us by their want of reason, and are therefore by the just appointment of the Creator subjected to us to kill or keep alive for our own uses; if so, then it remains that we understand that commandment simply of man.[84]

Augustine relies upon the Aristotelian construction of plants and animals as radically Other, in order to justify the biblical assertion that plants and animals exist for the sake of humans. The alleged superiority and preeminent position of humanity relies upon radical separation and alleged *dissociation* of humans from other beings. For Augustine, God created beings that are radically different and arranged in a natural hierarchy; therefore,

> The rational soul ought not to worship as its god those things which are placed below it in the order of nature, nor ought it to exalt as gods those things above which the true god has exalted it.[85]

Humans are exalted above the rest of nature for they alone of the earthly beings are thought to be in possession of the rational soul. The irrational-souled beings, the animals which have sentience, come next in Augustine's order of nature.

Without reason, sentience, or breath and therefore completely without soul in this religious hierarchy, plants again are rooted at the bottom.

St Thomas and The Philosopher

Aristotelian and biblical ideas of nature also converge in the philosophy of Thomas Aquinas (1225–1274 CE), who is regarded as the Christian tradition's greatest thinker.[86] His most influential works are undoubtedly the *Summa Theologica* and *Summa Contra Gentiles*, which have both had a considerable impact on many areas of Christian doctrine. Although Thomas clashed with Augustine over the relative merits of the Greek philosophers, he concurred with his predecessor in his treatment of plants.

In keeping with the observations of the ancient Greek scholars, and contemporary botanists such as Albertus Magnus, Aquinas deviated from the Bible in admitting the existence of life in plants. As Augustine had done nearly a thousand years before him, he purposefully retained the dissociation between plants and human beings by using Aristotle's tripartite division of life[87]:

> Thus the life of plants is said to consist in nourishment and generation; the life of animals in sensation and movement; and the life of men in their understanding and acting according to reason.[88]

Founded on a perceived lack of sensation and intellect, Aquinas solidifies the constructed inferiority of plants by extrapolating it to spiritual matters. Using the authoritative work of Aristotle, he restricts the possession of intellect or reason to human beings. He characterizes this capacity as the ability to know and understand, an ability which Aquinas considers to be "unrestricted" in human beings. The intellect can know all types of situations that are nonphysical, and therefore, Aquinas considers it to be ethereal in nature. All knowing, it is considered as superior.[89] Thomas considers this ethereal intellect to be the human soul, present in all the body parts, but befitting its incorporeal nature, immortal. Thus, in *Summa Theologica*, he famously asserts:

> The principle of intellectual activity, the soul of man, is an incorporeal and subsistent principle. For it is manifest that man can, thanks to intellect, know the natures of all things. . . . Therefore the intellectual principle, which is called mind or intellect, has an activity of its own in which the body is not involved.[90]

As only the intellectual faculty was incorporeal, only human souls fit the criteria for true immortal souls. This Thomistic position is extremely significant in the history of Christianity as it marks the break from the Aristotelian understanding of the corporeal soul and the adoption of the now commonly held idea of the immortal Christian soul.

To achieve this elevation of the human soul in *Summa Contra Gentiles*, Aquinas argues directly against the immortality of the animal soul. His case partly rests on the fact that the Bible equates the life of animal with the blood; therefore, unlike the intellectual life of man, the soul of the animal presumably requires the permanence of the body.[91] Another strand of Aquinas's argument is based on Aristotle's claims in *De Anima* that only the intellectual soul is incorruptible. If it is corruptible, the sensitive soul must be capable of change and death, and therefore, Aquinas concludes that humans and animals do not share the immortal soul.[92] Thus, for Aquinas, "This eliminates Plato's theory that the souls even of brute animals are immortal."[93]

Notably however, Aquinas does not construct an argument against the immortality of the plant soul. As the Aristotelian soul of plants is basically growth and reproduction, there isn't anything in plants that could logically survive death. Following Augustine, Aquinas regards plants as lacking in any type of soul, whether mortal or immortal.[94] This argument against the plant soul holds to this day—few people think of plants going to heaven when they die!

Although life was admitted to plants, for Aquinas, their spiritual inferiority helps justify their rightful exploitation by human beings. Following Augustine, Aquinas regards the use of plants not to involve killing. Without any type of soul, plants are not listed within the realm of human moral consideration as they are not capable of being killed or harmed in any way.[95] This constructed spiritual inferiority of plants helps demonstrate the natural superiority of human beings and justifies the biblical-Aristotelian position that plants exist for the sake of human beings. Echoing Aristotle, Aquinas states:

> The order of things is such that the imperfect are for the perfect, even as in the process of generation nature proceeds from imperfection to perfection. Hence it is that just as in the generation of a man there is first a living thing, then an animal, and lastly a man, so too things, like the plants, which merely have life, are all alike for animals, and all animals are for man.[96]

Furthering the argument, he quotes Augustine that "by a most just ordinance of the Creator, both their life and their death are subject to our use."[97] Plants are subject to human use because Aristotle declared them devoid of reason, sensation, and impulse. To Aquinas, this was a sign from God that they were "naturally enslaved and accommodated to the uses of others."[98]

In Christian writing, this backgrounding of plants can be read as a deliberate move to expand human claims on the natural world while avoiding moral consequences. Rather than arising from natural physiological biases, the rendering of plants as passive and radically different is a deliberate process of exclusion.[99] This is a process that can be identified in major streams of Western thinking and that comprises part of the contemporary cultural filter we use when viewing the constituents of the plant kingdom. Although Nash doubts the extent of the "cultural authority" of Christianity, historically it is clear that the biblical exclusion of plants from moral consideration and relationships of respect has formed part of the Western world's cultural attitudes toward plant life.[100] In contrast to this perspective of exclusion the following chapter will begin to evaluate non-Western sources that are orientated toward including plants within human moral consideration.

4

DEALING WITH SENTIENCE AND VIOLENCE IN HINDU, JAIN, AND BUDDHIST TEXTS

> All beings are fond of life, like pleasure, hate pain, shun destruction, like life, long to live. To all life is dear.[1]
> —Mahavira

The term *Hinduism* comprises a multitude of religious practices and philosophies, which although often grouped together, many scholars reject as being a unified or historical tradition.[2] Within Hindu philosophy itself, six separate and distinct schools are commonly delineated.[3] Faced with this diversity, it is impossible to treat Hinduism as a singular religion or indeed a singular philosophy, providing a singular perspective on the plant kingdom and a singular moral attitude toward it. Rather than attempt the impossible, in this chapter, I will instead simply put forward the notion that a number of Hindu religious scriptures (including the *Vedas*, *Upaniṣads*, *Bhagavadgītā*, and the *Mahābhārata*) offer specific examples of inclusive attitudes toward plants.[4] Whereas Callicott doubts the efficacy of a worldview that has been characterized as "world-denying," there is evidence in Hindu texts that plants are included within moral consideration.[5]

Within a broadly defined Indian philosophy, the notion of acting nonviolently toward plants first appears in (broadly defined) Hindu texts; however it is expressed most forcibly in the texts and practices of Jainism. In this chapter, I will survey a number of Jain sutras in order to explain how the theology of the Jain tradition incorporates the plant kingdom within an ethic of nonviolence.[6] Although Callicott also doubts the suitability of Jainism for environmental ethics because of its primary orientation toward human spiritual concerns and its renunciation of the world, Jainism is an important case study in a plant context because Jain spiritual practice (as well as theology) strongly incorporates plants into moral relationships.[7]

Whatever the orientation of the Jain tradition, it is clear that Jain scriptures are an excellent source for analyzing the points which lead to a recognition of plants as sentient, autonomous beings. Moreover, Jainism is an interesting case study not only because of its logical consistency and its recognition of individual presence, but also because of its focus on the enactment of small moral deeds aimed at reducing harm to others.

While positions of inclusion can be identified both within Hindu and Jain theology, it is not the aim of this chapter to delineate disrespectful and respectful treatments of plants along an East-West divide in the way that has previously been attempted with a wider environmental ethic.[8] While the idea of plants as passive, semi-alive entities appears to characterize Western philosophical and religious positions, it would be inaccurate to portray this treatment of plants as a purely Western phenomenon. Such a characterization of plants is also true of several major Buddhist schools.

This chapter will use the work of Buddhist scholar Lambert Schmithausen, who argues that in the early development of Buddhism, plants were ambiguously left as borderline beings in the context of human moral consideration. Subsequent dogmatic exclusion of plants from the realm of sentience and mind, and the concomitant loss of human-plant connection elicit comparisons with similar processes of exclusion in Western thought. This chapter aims to build upon the work of Schmithausen by positing that plants have also suffered deliberate exclusion in a number of Buddhist texts in order to avoid the uncomfortable inclusion of plants within Buddhist moral consideration.

Shared Substances and Plant Natures

Although studies of humans and plants abound in India, this case study differs by focussing on issues of human-plant connection, plant ontology, and moral consideration from a plant perspective.[9] The majority of existing studies on the relationships between humans and plants focus on the subject of sacred plants, which are a conspicuous feature of Indian ethnobotanical interaction. Plants are often considered sacred through a strong relationship with a deity or divine being.

To provide a few examples, bamboo (*Bambusa* spp.) is treated as sacred because it was used by the god Krishna for his flute. The sacred basil, *tulasi* or *tulsi* (*Ocimum tenuiflorum* L.) is also strongly associated with Krishna. Such associations are obviously powerful, for *tulasi* is commonly regarded as the most sacred plant in India.[10] Other instances of sacredness arise when deities reside in plants. The neem tree (*Azadirachta indica* A.Juss.) is regarded as the home of Shitala (goddess of smallpox), and the goddess Lakshmi is described as residing in the branches of the pipal tree (*Ficus religiosa* L).[11] In addition, tree deities—*devata* (male) or *devi* (female)—can inhabit both individual trees and groups of trees.[12]

While studies of sacred plants abound in India, they do not commonly emphasize substantial connections between plants and human beings. Yet there are instances in many ancient Indian texts where plants are not simply associated with deities. In many cases, plants are described as being formed from the body parts of a deity. The *Vamana Purana* notes that each species is cherished by its particular deity and lists over twenty species that have been created from divine body parts. *Tulasi* is considered to be the metamorphosed friend of Sri Radha.[13] The *padma,* or sacred lotus (*Nelumbo nucifera* Gaertn.) is described as originating from the navel of Prajapati. *Dhattura* (*Datura metel* L.) was formed from the heart of Mahesvara.[14] Common agricultural plants have also been generated from the bodies of deities. The *Matsya Purana* relates that caroway, rice, sugar, and pulses all originated from Vishnu.[15] In contrast to the drive toward difference and separation between the plant kingdom and the breath of God that is found in biblical passages, it appears that the Puranas regard plants as actually being *consubstantial* with the divine. In other Hindu texts, this idea of connection and fundamental relatedness extends from the divine to plants, rocks, animals, and human beings.

In particular, core texts in the influential Vedānta (*advaita*) school of Hindu philosophy contain passages which evoke a basic sense of connectivity between *all* beings. This is a common motif, which Bruce Lincoln regards as characteristic of the Indo-European tradition.[16] In his analysis of Indo-European mythology, Lincoln quotes the *Rig Veda*:

> The moon was born of his mind; of his eye, the sun was born;
> From his mouth, Indra and fire; from his breath, wind was born.
> From his navel there was the atmosphere; from his head, heaven was rolled together;
> From his feet, the earth, from his cardinal directions.[17]

In addition to the *Rig Veda*, the *Upaniṣads* also contain passages that clearly relate that *everything* manifests from and originates in Brahman—the omniscient, eternal divinity that both underpins and transcends the whole of universal reality. This is later echoed in a passage about Brahman from the *Mahābhārata*:

> Mountains are his bones, earth is the flesh, sea is the blood and sky is his abdomen.[18]

The *Chāndogya Upaniṣad* relates that "All this that we see in the world is Brahman"[19] and the presence of the supreme being in everything is also one of the main ideas of the hugely influential *Bhagavadgītā*. This notion of connection is succinctly summed up by Krishna who simply says "he resides in everywhere."[20] This is further expressed in the *Mahābhārata* in a passage addressed to Brahman which systematically lists all the worldly things that emerge from Brahman, including even the most weedy forms of plant life. As the passage says, "You are the thin creepers, you are the thicker creeping plants, you are all kinds of grass and you are the deciduous herbs."[21]

Deities, human beings, animals, plants, rocks, rivers, clouds, and air all derive from Brahman and so all are fundamentally connected.[22] In Advaita Vedānta philosophy, this manifests in the monistic idea that everything in its true essence is in reality a manifestation of Brahman. While keeping in mind Brockington's differentiation of the concept of personal deity from the absolute, there are similarities with the Christian notion that all beings share their creative origins in God.[23] However, there is a significant difference between these philosophies in the rendering of plant life. In the biblical text, plants and humans do not share in the same nature. But in Hindu texts, plants and human beings appear to share basically similar ontologies.

A number of influential Indian texts describe plants, animals, and humans all as *jiva* (living beings in possession of a personal self which undergoes rebirth—see next section). One of the most prominent of these texts is the *Bhagavadgītā*, which states clearly that all living beings are in possession of *jiva*.[24] The basic nature of this *jiva* appears to be the same for all ensouled beings. For any living being, sensation and awareness are essential qualities. It could be argued that there is also no concept of life without these faculties. Indeed in the *Mahābhārata* it is stated that "sentience is the mark of life," a point reiterated in the *Bhagavadgītā*:[25]

> The living entity enjoys sense pleasures using six sensory faculties of hearing, touch, sight, taste, smell, and mind.[26]

As an embodied soul, as a *jiva*, the plant is recognized as sentient, volitional, and aware. The *Bhagavadgītā* states that the senses and the mind are associated with the soul, and as the body dies, the soul takes the senses into another incarnation.[27] Here there could be doubt about whether plants are recognized as living beings. However, by referring to *jiva* as both *mobile* and *immobile* beings, it is clear that the popular Indian epic the *Mahābhārata* regards this autonomous self to be found in plants.[30] Indeed, it is clearly stated: "Those seeds of grains they call rice and so forth, they are all alive. . . ."[29]

It is important to note that the relevant passages from the *Mahābhārata* do not parallel Aristotle by portraying this plant soul as radically different and inferior to the human. Instead of hyperseparating the plant and the human, it is clear that from careful observations, plants are beings that are very similar in nature to humans. Certainly plants possess sensation and awareness:

> They [plants] sicken and dry up. That shows they have perception of touch. Through sound of wind and fire and thunder, their fruits and flowers drop down. Sound is perceived through the ear. Trees have, therefore, ears and do hear. A creeper winds round a tree and goes about all its sides. A blind thing cannot find its way. For this reason it is evident that trees have vision. Then again trees recover vigour and put forth flowers in consequence of odours, good and bad, of the sacred perfume of diverse kinds of *dhupas*. It is plain that trees have scent. They drink water by their roots. They catch diseases of diverse kinds. Those diseases again are cured by different operations. From this it is evident that trees have perceptions of taste. As one can suck up water through a bent lotus-stalk, trees also, with the aid of the wind, drink through their roots. They are susceptible to pleasure and pain, and grow when cut or lopped off. From these circumstances I see that trees have life. They are not inanimate.[30]

This description of plant life in the *Mahābhārata* is based upon an understanding of common origins and shared substance, and a drive toward connection rather than separation. It is also dependent upon an acute awareness of (and emphasis on) similarities in the life histories of plants, animals, and humans. Appropriately these shared faculties are not evaluated from a zoological perspective, and the acceptance of these human-plant connectivities leads to the natural conclusion that plants are living, autonomous entities. Taking the case even further, the extremely influential Vedāntic text, the *Yogavāsistha*, portrays the living plant as not only sentient but also self aware. In a remarkable passage it is stated:

> As a tree perceives in itself the growth of the leaves, fruits and flowers from its body; so I beheld all these arising in myself.[31]

Hindu texts are not univocal, however, and this perception of plants in the influential *Bhagavadgītā* is interestingly and directly contradicted by a passage in the *Vishnu Purana*. This later Puranic text disputes the notion that plants are perceptive and, in common with the biblical understanding of plants, briefly describes the plant kingdom as nonsentient. Although it describes the divine origin of plants, and their emergence from the hair on the body of Brahma, the whole plant kingdom is portrayed as:

> The fivefold (immovable world), without intellect or reflection, void of perception or sensation, incapable of feeling, and destitute of motion.[32]

Despite recognizing that plants are alive (and, thus, are *jiva*) and have a divine essence, the *Vishnu Purana* denies them any sensory or mental qualities. Such an interpretation of living beings without these faculties appears to contradict the very nature of the embodied soul as set out in the *Bhagavadgītā*. Nevertheless it demonstrates the existence of zoocentrism even within more inclusive perspectives, a point which will be taken up in the section on Jainism.

Heterarchical Connections

The treatment of plants as ensouled beings is significant for it provides a strong point of connection between the plant and human worlds. It is supplemented by passages in the *Mahābhārata* which recognize that the worlds of plants, humans, and animals interpenetrate at death through karma and rebirth.[33] Although the position of plants in the cycle of rebirth is often unclear (with plants often left out of many interpretations of Hindu rebirth), the *Mahābhārata* states that "*All creatures, stupefied, in a consequence of ignorance, by the attributes of goodness and passion and darkness are continually revolving like a wheel.*"[34] In the *Mahābhārata*, it is clear that *all* creatures participate in this cycle (known as *samsara*). As in the attribution of *jiva,* there is no discrimination between mobile and immobile beings; no discrimination between animals and plants. All the ensouled existences interpenetrate. Even though humans may have greater potential for enlightenment in this life, our current positions and current lives in the plant, animal, and human realms are interchangeable at death.

This interpretation of *samsaric* existence is not limited to the *Mahābhārata*. It first appears in the *Upaniṣads*, where the doctrines of karma and rebirth are first enunciated. The first mention of karma occurs in the *Brahdāranyaka Upaniṣad*, when the sage Yajnavalkya is asked by the scholar Artabhaga what happens to a person after they have died.[35] It is indicated that there are two paths, which can be taken, the path which leads to a final existence or the path which leads to a rebirth in the world. This second path into another birth is described in very ecological terms, a process which very much includes the vegetation.[36] Each cycle begins with death and the transformation of the deceased into smoke through burning on the funeral pyre. After a brief sojourn in the realm of the gods, the smoke then forms part of the sky. It merges with the clouds and falls to earth again as rain. Here, the individual is reborn as vegetation, which in turn is then transformed into other types of living beings, perpetuating the cycle.

The potential to be reborn as a plant also appears in a passage in the *Chāndogya Upaniṣad*. Again a very physical process is described, and the dead person is once more transformed into smoke by the funeral flames. The smoke again merges with the rainclouds in the sky and

> Having become cloud, it becomes rain cloud; having become rain cloud it rains. Here they are born as rice and barley, herbs and trees,

sesame and black beans. It is very hard to escape from this. But whenever someone eats the food and emits seed, once again [it] comes into being.[37]

Although the *Chāndogya Upaniṣad* portrays plant life as a realm of unfortunate birth, from which it is difficult to escape, plants are not radically separated from the human and animal realms. They are connected in a web of life and death. After death, it is apparent in the *Upaniṣads* that it is possible for a human soul to be reincarnated in the plant kingdom—in herbs, trees, and even cultivated crops. As plants are fully ensouled, they also have the potential to be born as human beings and to gain liberation. As the soul can manifest in all types of beings, here the *Upaniṣads* recognize that plants are spiritually and physically interdependent with the rest of the natural world.

Interestingly, in the well-known text *Manu Smrti*, the possession of sentience in plants is explicitly linked with the capacity for suffering. All plants including trees, creeping plants, grasses, and herbs are said to "possess internal consciousness and experience pleasure and pain."[38] Logically, as *samsaric* beings, plants must be able to undergo suffering, as suffering permeates the *samsaric* world. As ensouled, sentient, beings, in the same way as humans, plants strive to escape this *samsaric* suffering.

This connective basis also appears in an extract from the *Mahābhārata*, which states that "animals, human beings, trees and herbs, all wish for the attainment of heaven."[39] As participants in the cycle of suffering, relationships with plants ideally should be based upon respectful and compassionate action, both out of "karmic selfishness" and out of respect for others. Ideally they should be based on nonviolence (*ahimsa*), for "nonviolence is the highest Law."[40] Throughout the *Mahābhārata*, there is a stress on kindness toward all creatures; virtuous people "are always compassionate to *all* creatures and devoted to nonviolence."[41]

Devaraja also draws attention to a very significant reference to *ahimsa* which appears in the *Yoga Sutra of Patanjali*.[42] With particular resonance for our age of extinction, the practice of nonviolence is regarded as essential for the continued existence of other living beings. Here the sutra states:

> Along the bank of the great river of compassion lie the different creeds that are comparable to grass, sprouts, (plants), etc., if that river (of compassion) goes dry, how long can the latter prosper?[43]

In the same way as they need water, in this passage it appears the different types of plants need our compassion in order to flourish. Although the ideal behavior is to exercise *ahimsa* in all dealings with living beings, it is self evident that violence must be done to plants for humans to survive. Commonly much of this ideal of *ahimsa* is directed toward the avoidance of harm to animals (resulting in a commitment to vegetarianism), however there is recognition that any form of eating involves harming plants and taking their lives.[44]

Lance Nelson has questioned the validity of the claim that *ahimsa* is an ecological virtue by pointing out that references to *ahimsa* in the *Bhagavadgītā* are "articulated for the most part out of concern for the private karmic well-being of the [human] actor. Compassion for others, while a significant factor, is secondary."[45] Connected to this is Nelson's argument that the philosophical representation of physical reality as "non-real" or transient means that "the physical, including the empirical existence of other beings, does not matter."[46] Although these are significant doubts, a couple of important points can be made in response.

Firstly, the claim that the physical existence of other beings does not matter needs to be put in context. In the *Bhagavadgītā,* references to the transience of the world are spiritual teachings explaining the absolute state of existence, recognizable by persons free of ego and attachment.[47] It is important to point out that such metaphysical concepts are not intended to provide a framework for human moral action. Indeed the text emphasizes that moral action in this embodied life is very important, adding weight to the claim that corporeal existence "matters." Often, exhortations to moral action arise from a recognition of the connections and similarities in the lives of beings.[48] From this basis, it can be argued that even if the primary motivation of nonviolence is concerned with *self,* the action of nonharm is also concerned with *others* because it is recognized that not being harmed matters to other beings. This is something of a *dialogue* of nonviolence, which is part of the relationship between self and other.

Jainism and Living Plants

Even though notions of plant sentience and nonviolence are embedded within these texts, it would be misleading to assert that they are generally at the forefront of religious practice within the broad Hindu tradition. For a discussion of a tradition that clearly and definitively embeds teachings on nonviolence and plant sentience into its daily practice, it is necessary turn to the Jain religion, which emerged from the philosophical heritage of ancient India in the sixth century BCE.

Jainism contains ideas similar to the Hindu notions that are relevant to the study of the relationships between plants and people.[49] However, rather than a disparate smattering of references to plant ontology and ethics across a number of texts, Jainism is notable as a defined tradition with a cohesive, explicit stance on the nature of plants and human-plant ethics. More importantly, rather than remaining on the level of the philosophical, the Jain teachings on the autonomy and sentience of plants are remarkable for they are at the forefront of Jain practice. In Jainism, it is very important to incorporate lifestyle measures that mitigate the destruction of plant life as far as possible.

In the same way as the passage in the texts brought together under the loose umbrella of *Hinduism,* Jainism emphasizes connection and similarity between beings rather than separation and distance. Once more, humans and plants are recognized as possessing a shared ontology. In Jainism, the world is divided into two broad categories, the living and the nonliving. The living world of Jainism contains countless life souls (*jiva*), and plants are recognized as *jivatthikayal* (living or ensouled bodies). In contrast to the Bible, the earliest Jain sutra recognizes plants as living beings based upon the ability to flourish. In the *Acaranga Sutra*, a famous statement from the Tirthankara Mahavira encapsulates this reasoning:

> As the nature of this (i.e., men) is to be born and to grow old, so is the nature of that (i.e., plants) to be born and to grow old; as this has reason, so that has reason[50]; as this falls sick when cut, so that falls sick when cut; as this needs food, so that needs food; as this will decay, so that will decay; as this is not eternal, so that is not eternal; as this takes increment, so that takes increment; as this is changing, so that is changing.[51]

For Mahavira, there is a clear connection between the lives of human beings and the existence of plants. In the terminology of philosopher Chakravarthi RamPrasad, Mahavira is seeking out *affinity* with other beings in the face of clear and obvious alterity.[52] As will be explored in the forthcoming discussion, this search for affinity is a strong component of the fundamental Jaina ethic of *ahimsa*.[53] In the above quotation from the *Acaranga Sutra*, it is clear that there is the recognition of ontological affinity. This recognition is not simply based upon woolly wishes for nonviolence, but on logical consistency. As plants seek to thrive in similar ways to humans, (seeking out food and maintaining integrity) and also undergo death, life is acknowledged in plants. A further passage in the *Acaranga Sutra* uncovers how this recognition of vitality is brought about:

> Thoroughly knowing the earth bodies and water bodies and fire bodies and wind bodies, the lichens, seeds and sprouts, he comprehended that they are, *if narrowly inspected*, imbued with life.[54]

As plants differ greatly in form and process from humans and animals, this evokes the need for an acute state of mindfulness when examining the possibility of life in a plant. Mahavira suggests that only through a thorough and sensitive inspection of earthly creatures will human beings realize their living nature. Such close inspection of slower, more subtle workings also yields the knowledge that seeds, sprouts, and lichens are living beings too. Each is a *jiva*.

Vital processes such as respiration may not be directly observed, but from more noticeable changes in gross morphology and behavior, it is inferred that they do occur. Although it is true that plants cannot be observed to breathe, Mahavira infers that "these living beings with one sense also inhale and exhale, breathe in and breathe out."[55] Such awareness is remarkable; both in light of contemporary knowledge of plant respiration and also in contrast with the biblical insistence that plants lack the breath of life.

Affinity and Otherness

As an embodied *jiva*, a plant has the same basic ontology as a human being. Jain texts also make it clear that the nature of this soul is the same in all beings. The nature of *jiva* is awareness. The *Tattvarta Sutra* states, "Sentience is the defining characteristic of the Soul."[56] The fact that plants, animals and humans each possess this sentient soul strengthens the notion of connectivity.[57] Again as embodied souls, the realms of plant, animal and human interpenetrate after death. In the next life a human being could manifest as an animal or a plant and vice versa. Mahavira himself describes how trees can be born as human beings and how like human beings they have the potential to attain enlightenment.[58] Therefore, across the wide variety of life forms on the Earth, in the words of Cort, Jainism posits "an interdependent continuity of existence."[59] This is an existence of *multiplicity* and *affinity*.[60]

Plants also share affinities through many of the faculties that they possess in life. From close observation of their life history, plants are regarded as possessing sensation as well as the attributes of flourishing beings; the desire for food and reproduction, an awareness and orientation toward the future, and an aversion to physical harm.[61] In addition, plants are recognized in Jain literature as being subject to a range of emotions produced by karma—such as passion, anger, pride, deceit, and greed.[62]

Plants and humans have the same essential nature, but they have differences in their faculties in life. Jainism does not position the parts of the natural world in an anthropocentric hierarchy of absolute worth, nor deny the autonomy of other living beings. However Jainism does describe a *relative*, nonviolent hierarchy in the natural world based upon the development of the soul's potential for sentience in each living being.

Jainism first classifies living beings according to motility.[63] Plants along with earth and water are classified as immobile beings.[64] The immobile beings are regarded as only possessing one sense (touch), and are often referred to as one-sensed beings throughout the Jain literature.⁶⁵ The mobile beings, such as animals and humans, are regarded as possessing a range of the senses (from two to five)—with a separate category of five-sensed beings that also possess rationality. Larger animals such as fish, birds, quadrupeds, and humans are said to have a physical basis for mind and as such are regarded as intelligent beings capable of memory and reasoned thought.⁶⁶ Humans are differentiated from the rest of the living world as the soul can only be liberated when it is in the human form. Crucially, the defining aspect of the human sphere of life is not constructed in terms of dominant superiority, but in the ability to act morally and to practice *ahimsa*.⁶⁷

Thus, in Jainism's classification of the natural world, plant activities such as the acquisition of food are considered as survival instincts rather than as intelligent, reasoned behavior involved in "judging objects and situations that arise in the wake of specific enquiry."⁶⁸ In this rejection of intelligence and reasoning in plants, Jainism differs markedly from other traditions that recognize sentience and autonomy in the plant kingdom. In particular this position differs from the description of minded plants in texts such as the *Yogavāsistha* and the *Mahābhārata*.

From the description set down in the *Tattvarta Sutra*, this denial may be partly based upon zoocentric reasoning similar to that which appears in Aristotle. As plants do not obviously possess a brain in the same way as animals, the sutra states that they do not have a physical basis for mind. However, it fails to consider that plants may have their own anatomy for reasoning.⁶⁹

Despite the undervaluation of plant capabilities, plants are not constructed as a radical and inferior Other in relation to the human. Although a slight zoocentrism can be detected, Jainism does not engage in a process of backgrounding, as a preparation for the human domination of the natural world. Jainism's main orientation is to build moral consideration based upon the affinities that connect living beings in all their diverse morphologies and modes of life. As Ram-Prasad suggests, Jaina ethics is founded upon a search for affinity in the face of clearly discernible and identifiable otherness.⁷⁰ The engaged nonviolence of the Jainas is built upon a consideration of the affinities that connect different modes of existence.⁷¹

Nonviolence Toward Plants

As beings possessed of *jiva*, plants have a shared capacity for awareness, autonomy, and volition—capacities in common with all other living beings. It is this view of nature, which is able to flourish, that determines the Jain action toward the plant kingdom. In a famous passage from the *Acaranga Sutra*, Mahavira states:

> *All* beings are fond of life, like pleasure, hate pain, shun destruction, like life, long to live. To all life is dear.[72]

The fact that each living being clearly values its own life, seeks to preserve its integrity, and seeks to blossom and reproduce is extremely significant in the Jain tradition. Violating this autonomy and causing harm to living beings is a negative act that results in the accumulation of karma.[73] Therefore, for the Jain, the recognition that the world teems with a multitude of sentient, self-aware and self-cherishing forms of life requires the adoption of a rigorous practice of nonviolence. Here, nonviolence need not simply be negatively understood as a selfish human tactic for reducing the accumulation of karma. It can also be determined positively as the cessation of human harm and violence which "creates the space in which the Other can flourish."[74]

Traditionally however, in the Jain tradition, *himsa* is defined as "the deprivation through carelessness of any one or more of the vitalities of the soul."[75] The Jain sutras place great emphasis on the avoidance of harmful actions toward others. The *Acaranga Sutra* states:

> All sorts of living beings should not be slain, nor treated with violence, nor abused, nor tormented, nor driven away.[76]

Unlike in the Western traditions already surveyed, moral consideration does not reside solely with the human. All beings that are capable of being harmed are included within the moral sphere. It could be said that the possession of life, autonomy, integrity, and volition marks the moral boundary.[77] As living beings, plants are morally considerable, and the effects of human action on plant welfare must be seriously considered and evaluated. Here is a crucial exercise in logical consistency. Rather than play down, or deny the fact that plants are alive, the Jain tradition requires its adherents to act in full possession of this knowledge. Jains must recognize the sentient nature of plant life and ideally must seek to minimize the harm which is done to plants in order to let them live.

For monks and nuns, the ethical practices, which concern plants, are encapsulated in the first great vow, in which they completely renounce both the killing of all living beings and causing others to kill living beings. According to the *Acaranga Sutra,* the renunciant and the person possessed of wisdom must display the knowledge that plants possess sentient life and must accordingly avoid harming them.[78] One of the consequences of this stance is some severe dietary restrictions. As well as a commitment to vegetarianism, Jain monastics are subject to ethical restrictions in their consumption of plants. The *Acaranga Sutra* states that monks and nuns should not accept any raw plant foods as their consumption will involve direct killing. The sutra forbids the consumption of any raw plants, roots such as ginger; raw fruits and berries; raw shoots; raw sprouts; raw sugar cane; raw bulb-like parts such as garlic; and raw grains.[79]

For lay people, these rules are not as strict. Rather than avoid any act of violence at all, the lay person must instead try to avoid any unnecessary destruction of life.[80] This can be as simple as avoiding the waste of paper and plant foods. Even for the monk, the ideal of nonharm to all beings is simply unattainable, for any life must involve the killing of plants by someone. The point is that the use of plants is recognized to involve killing and that this is minimized wherever possible. Avoiding all harm to living beings is an ideal to be striven for.[81] As Callicott says of ethics in general, they are "never perfectly realized on a collective social scale and very rarely on an individual scale."[82]

One interpretation of the general Jain understanding is that plants, animals, and humans have needs which often conflict. Despite the ideal of nonviolence, plants must be eaten for other living beings to remain alive. In such cases, Wiley has put forward the idea that there is a sliding scale of suffering based upon the number of senses and levels of awareness; therefore, the karma consequences of harming a plant are less than causing harm to an animal or a human being.[83] However, the *Acaranga Sutra* is emphatic that the wise person recognizes the equality of all living beings.[84] And it even states that "all kinds of living beings feel the same pain and agony."[85] Therefore, regardless of the type of being, it is imperative to minimize violence wherever possible.[86]

For a devout Jain, actions must be taken to mitigate violence. These can range from starvation (in the Digambara tradition as an act of moral defense for other beings) to the simple application of mindfulness and carefulness in the gathering of human food. As Shilapi notes of six hungry friends:

> As they ran to the tree the first man said "Let's cut the tree down and get the fruit." The second one said. "Don't cut the whole tree down, cut off a whole branch instead." The third friend said, "Why do we need a big branch?" The fourth friend said, "We do not need to cut the branches, let's just climb up and get the bunches of fruit." The fifth friend said, "Why pick that much fruit and waste it? Just pick the fruit we need to eat." The sixth friend said quietly, "There is plenty of good fruit on the ground, so let's just eat that first."[87]

The Jain tradition demonstrates that human beings can do a great deal to mitigate the conflicts that arise with others. This can be achieved through a simple and honest appraisal of what our real needs are. As plants are autonomous beings, it means that we should ensure that not all of our relationships with plants are instrumental. To lessen conflict, we should reduce human claims.[88] Thus, Mahavira relates:

> A monk or a nun, seeing big trees in parks, on hills, or in woods, should not speak about them in this way, "These [trees] are fit for palaces, gates, houses, benches, bolts, boats, buckets, stools, trays, ploughs . . . [instead they] should speak about them in this way, "These trees are noble, high and round, big; they have many branches, extended branches, they are very magnificent. . . ."[89]

Concentrating solely on the instrumental value of plants would be a denial of their affinity with humans as equally valid locations of being. This exclusion would violate the Jaina ethic of nonviolence.[90]

Buddhism and Backgrounded Beings

In a recent review of Buddhist ecophilosophy, Donald Swearer has characterized five overlapping categories of scholarship concerning Buddhism and the environment, the proponents of which he terms "eco-apologists, eco-critics, eco-constructivists, eco-ethicists, and eco-contextualists."[91] The delineation of these contrasting stances within Buddhist environmental discourse is useful for the purposes of positioning the remaining sections of this chapter on human-plant relationships.

Buddhist eco-apologists have repeatedly claimed a basis for Buddhist environmental ethics in the Buddhist worldview of *paticca samuppada*, (the interdependent co-arising of all phenomena), *anatta* (no self), and *sunnata* (emptiness).[92] Extrapolated and applied to a plant context, such scholarship claims that Buddhism maintains an ethical relationship with the plant kingdom on the basis of human-plant interdependence. According to eco-apologists, it appears that Buddhism also emphasizes connection and affinity between plants and humans rather than separation and distance.

Such notions, however, have been criticized directly by Buddhist virtue ethicists Cooper and James, who have rejected the oft-recited notion that Buddhist teachings on *paticca samuppada*, *anatta*, and *sunnata* are appropriately applied to human interactions with the biosphere. They maintain:

> The doctrine of conditioned arising in its entirely general form—"this arising, that arises. . . . This ceasing, that ceases" (M115)—is treated by the authors cited [eco-apologists] as the claim that all events and processes are causally connected. . . . It becomes only interesting when the causal connections are spelt out in detail, something that is done only in cases on *human psychology*.[93]

A counter argument could be constructed by pointing out that although not originally intended for the purpose, such notions of interdependence could still be directed toward respectful relationships with nonhuman others. Although any attempt to find the foundations for respect are to be applauded, relying solely on such notions of interdependence is not always adequate in the moral sphere. As a stand alone concept, interdependence does not take us far enough into the moral domain. It gives us no framework for discerning between moral concern for rainforests and/or nuclear dumps. As ecocritic Ian Harris makes clear, it also does not make clear which entities we should consider as moral subjects in a time of ecological degradation.[94] Indeed, for this reason, Cooper and James argue that interdependence provides "a very fragile basis on which to erect any substantial account of the empirical relationship between human beings and the rest of the living world."[95]

Significantly, in a botanical context, notions of interdependence are coexistent with the moral exclusion of plants. Fundamental Buddhist doctrine (found in the Pali canon) hyperseparates plants and human beings by not including plants within the realm of sentient beings. By doing so, Buddhist texts strip plants of the attributes (of sentience) that Buddhists regard as circumscribing moral consideration.

With its parallels in Western philosophies, such exclusion of plants is particularly visible in Tibetan Buddhism. The Bhavachakra, the Tibetan Buddhist wheel of life, represents the processes of life, death, and rebirth in the continuous cycle of *samsara*.[96] A remarkable feature of the wheel of life is that while life is recognized in celestial beings, it is denied to plants. Plants are conspicuously absent from the wheel of sentient life.

The representation of plants in the Bhavachakra can be traced to the *Majjhima Nikāya*, a constituent text of the Pali canon and a fundamental text in Theravada Buddhism. In a short passage, the Buddha relates the possibilities of rebirth for a human being after death. He speaks of five different destinations which are "hell, the animal realm, the realm of ghosts, human beings and gods."[97] In contrast to the account of rebirth as it first appeared in the *Upanisads*, in the *Majjhima Nikāya*, plants are not considered a possible destination for rebirth. Neither are they included within the four types of observed birth which are recorded as "egg-born generation, womb-born generation, moisture born generation, and spontaneous generation."[98] This rendering of plants as insentient and passive sharply contrasts with ideas in Hindu texts and the Jain tradition.

Buddhism generally describes living experience as being composed of five aggregates (*skandhas*). The first of these is form (*rupa*) that refers to the domain of material existence. The remaining four aggregates are sensations (*vedana*), perceptions (*samjna*), psychic constructs (*samskara*), and consciousness (*vijnana*).[99] In this understanding, sensation, perception, psychic constructs, and consciousness combine to form the notion of sentience, but sentience is also a hallmark of life. Therefore, as plants are perceived to be nonsentient, their status as fully living beings is also called into question.

Reversal of Plant Sentience

The broad Buddhist tradition provides an interesting case to study the process of plant exclusion as it appears to be a reversal from an earlier position of inclusion. Schmithausen's *Problem of the Sentience of Plants in Earliest Buddhism* offers a detailed analysis of whether the doctrine of plant nonsentience was the earliest position of the Buddhist tradition. It draws attention to a body of evidence that supports the idea that at one time plants were considered by early Buddhists to be living beings capable of sensation. Fundamental to this idea is a passage from the *Patimokkhasutta*, the basic code of monastic discipline, for Theravada which states:

> If [a monk or nun] is ruthless with regard to plants, this is an offence to be atoned.[100]

Although in this short rule there are no direct references to living or sentient (*pana*) beings in the Pali version of this text, there are introductions which explain the rules.[101] In the case above, there are two short stories given to demonstrate why plants should not be injured.[102] One of these stories relates that when Buddhist monks felled trees, the local people disapproved of these actions because they regarded the trees as sentient beings.[103] Although this implies that the monks did not share the same beliefs, the text associates the felling of trees with the killing of animals.[104] Such ideas are also found in the *Sutta Pitaka*, where a commitment to neither kill nor harm seeds and plants forms part of the moral life of the monk.[105]

However, Schmithausen considers that this is evidence is not particularly explicit. The inclusion of a rule in the monk's moral code may relate solely to matters of ascetic decorum than morality.[106] He notes a more interesting passage in the *Sutta Pitaka*, which as an earlier text than the *Nikāyas* and *Pitakas* may indeed be more representative of the earliest traditions of Buddhism. In the *Sutta Pitaka*, the first precept for lay people is that they should abstain from killing and acting violently toward both *mobile* and *stationary* animate beings (which as in Jainism is taken to mean plants).[107] This rule echoes the position of the Jains by not restricting the notion of killing to animals. As in the Jain tradition, for laypeople, it is also formulated as an ethical ideal to be striven for.[108]

Schmithausen provides more evidence for such a position in a passage from the *Vinaya Pitaka*, the monastic rules for monks and nuns. In this passage, the Buddha chastises monks for cutting palmyra leaves (*Borassus flabellifer* L.) and bamboo leaves for use as sandals, again on the basis that the local people regard them as *one-sensed* living beings.[109] This prohibition on cutting plants could again be aimed at not offending the beliefs of the local people.[110] However, the passage suggests that the cutting of the leaves may actually be causing pain to these one-sensed beings, who express this by withering up.[111]

DEALING WITH SENTIENCE AND VIOLENCE   89

In a similar position to Schmithausen, Buddhist scholar Ellison Findly asserts that the early Buddhist tradition regarded plants as one-sensed beings, which possess the faculty of touch, finding evidence from the descriptions of plants in the Pali canon to support his claim. Plants are observed to be sensitive to the environment around them.[112] They prosper in warm sunlight and die back in response to the cold of winter. Their well-being is noticeably improved in response to well watered soil, rather than dry, cracked earth. Furthermore, plant growth involves the extension of roots and branches and the negotiation of stones and boulders as well as winds and rains, feats which must involve the use of sensing.[113] And in the case of seeds, it is certain that the early texts are aware of the need for fertile soil and continuous water and nourishment for the seeds to germinate and grow.[114]

THE QUESTION OF USE

This early material reveals that early Buddhism may have considered plants as living beings possessed of some measure of sensory awareness. Although this position can be doubted on the sparseness of supporting material, what is certain is that Theravada Buddhist monastics did not (and do not) face the same strict censures on the use of plants for foods as their Jain contemporaries. This issue of consuming plants may go some way to explaining how plants came to be regarded as insentient beings, devoid of sensation and mind.

There is obvious conflict between the need to use plants for food, clothes, and shelter and the Buddhist teachings on suffering and nonviolence. If plants are considered as sentient beings, great practical difficulties emerge when attempting to live by the first precept of Buddhism not to take life, or to live without causing suffering to others. It is a fundamental fact of embodied life that some beings must be harmed and killed in order to sustain others. Human subsistence relies entirely, whether directly or indirectly, on killing plants.[115]

Although very early texts such as the *Sutta Nipata* regard plants as animate, later compositions increasingly and deliberately avoid making explicit statements on their nature. Schmithausen argues that in order to avoid the more extreme aspects of Jain philosophy, the earliest formulations of Buddhism attempted to reconcile the need to eat with the principle of *ahimsa*, by rendering the question of plant sentience deliberately ambiguous. Plants were deliberately left as a borderline case in order to avoid extreme acts of asceticism (such as self starvation) and to focus the minds of both monastics and laity on the issue of training the mind. Explicit statements on the nature of plants were put to one side to "deliberately avoid arousing in lay people qualms in connection with a moderate utilization of plants for food and other basic needs."[116]

In the context of this study, this can be seen as a subtle, yet deliberate backgrounding of the plant kingdom in connection with human use. Rather than recognize that plants are autonomous, aware, and perceptive (and so recognize that all eating involves killing), it appears that early Buddhism quietly played down the killing and violence done to plants. Buddhism thereby placed plants outside the realm of moral consideration (principally in terms of the first precept of *ahimsa*). The rejection of sentience also excludes plants from that group of beings who are the appropriate recipients of virtuous actions.[117]

An Insentient Doctrine of Zoocentrism

From the early position of subtly, yet deliberately, leaving plants as borderline beings, the insentience of plants eventually became enshrined in Buddhist doctrine. Taking the tradition as a broad whole, Schmithausen asserts:

> Buddhism fell prey to the desire for an unambiguous theoretical position, which amounted to plants being virtually, and in the end, at least on the doctrinal level, explicitly, excluded from the range of sentient beings.[118]

In contrast to the Jain tradition, in the Buddhist doctrine that is relevant to the natural world, plants are radically separated from the human being by constructing and emphasizing discontinuity. Again, as in the case of Western philosophical and religious texts, this lack of affinity is stressed heavily in a number of Buddhist texts and is justified by evaluating plant life using zoocentric criteria.

In this drive toward disconnection, a common scholarly tactic for rejecting sentience is comparing and associating the behavior of plants with insentient things. In the *Madhyamakahrdaya* (an important text of the Madhyamaka School) when plant sentience is debated because of the evidence that plants grow, fall sick, and produce offspring just like animals, the riposte relies solely upon comparing plants with nonliving objects.[119] In reply to the argument that plants must be sentient because they need to be treated with medicine to recover from illness, the text replies that nonliving things such as "spoilt liquor or defiled gold are 'cured' by means of certain ingredients."[120] Moreover, the processes of growth, disease, and reproduction, which plants share with other living creatures, are compared to processes of expansion and contraction in nonliving things such as salt, crystals, and jewels.[121]

From a range of texts, Schmithausen lists a further set of reasons employed by a variety of Buddhist scholars many of whom are influential in the Tibetan tradition for separating plants from other living beings and denying their sentient status.[122] There are eight main reasons, which I have summarized for brevity:

1. Plants lack autonomous motion or locomotion.
2. Plants lack body heat.
3. Plants do not perceptibly breathe as do men and animals.
4. Plants do not get perceptibly tired.
5. Plants do not open or close their eyes, i.e., referring to the opening and closing of eyelids showing sleeping and waking and the changes of consciousness that these relate to.
6. Plants can regrow branches and stems which are cut, but animals cannot regrow limbs.
7. Plants do not answer when spoken to.
8. Plants do not perceptibly move when violently injured, from which it is deduced that they do not feel pain or pleasure.

The philosophical orientation of such reasoning makes it clear that the *established* doctrine is that plants are insentient. From this default stance of exclusion, it is clear that the criteria for sentience are all zoocentrically biased. The arguments employed against plant sentience are simply aimed at *justifying* the position established in the *Nikāyas*. Unlike in the earlier Hindu texts, in these important Buddhist scriptures, there is no emphasis on the shared origins and many points of connection between humans and plants. Human and animal characteristics remain the benchmark for discussions of plant sentience.

However zoocentrically biased this treatment may be, such arguments have helped establish plants as nonsentient beings in almost all Buddhist traditions.[123] On this basis, plants are also considered to lack mental awareness or consciousness. Such a pyramid of increasing exclusion can be identified in the Tibetan word for sentient being, *sem chen* (a being with a mind).[124] Such a deepening of exclusion has parallels with the modern Cartesian treatment of mentality and continues to affect contemporary Tibetan Buddhist thought. As a Lama of the Kagyu tradition states, sentient beings are "those with consciousness like ourselves, and other beings who experience happiness or unhappiness, pain or pleasure."[125]

In Theravada and Tibetan Buddhist metaphysics, plants—who are construed to be devoid of sentience—are positioned outside of the cycle of birth and death; whereas in both Hindu and Jain texts, plants play an active part in this process. Importantly, as plants are considered neither living nor sentient, they are traditionally viewed as being unable to undergo suffering. Within Buddhist schools, plants "are not generally considered sentient, and so are not protected under the first precept"; the precept that exhorts all practitioners to refrain from killing (sometimes interpreted as harming) sentient beings.[126] In general it could be argued that this exclusion of plants from sentience and from the protection of the first precept is also an exclusion of plants from moral regard. Such a position is exemplified by a contemporary Tibetan Buddhist Lama who reasons:

> Plants do not hold a consciousness, an individual consciousness. So when you cut a salad, a flower, a plant, you do not kill a being. This should be clear. Only in some particular cases, that some big trees have a spirit consciousness in them. In such case, it's not a being that has taken rebirth as a tree, but as a spirit who will be "linked" into a tree, as it could be into a house, a rock, or whatever else. The vegetal realm does not exist as a possible realm of rebirth. Thus, to cut a plant is not killing, by any way.[127]

In Tibetan Buddhism, therefore, the idea that plants are nonsentient is not confined to ancient scriptural passages; it is a commonly taught doctrine. The Dalai Lama himself is prepared to grant sensory awareness and consciousness to an amoeba, but is unsure of whether plants are sentient.[128] In the Mahayana teachings, this question is extremely important because the *Bodhisattva* primarily directs compassion toward sentient beings. The Dalai Lama however keeps a characteristically open mind on this issue and admits that plants *could* be sentient. He recognizes the "biological life" of plants and is open to scientific evidence overturning Buddhist claims.[129] In this respect, the abundant scientific evidence that plants possess sensory awareness (see Chapter 7) could help reverse the denial of plant sentience in Tibetan Buddhism.[130]

Does Moral Inclusion Really Matter?

Although they accept that plants are placed outside the realm of moral consideration, Cooper and James claim that certain Buddhist virtue ethics extend to beings that the tradition generally considers nonsentient. While admitting that plants are not regarded as appropriate recipients of key Buddhist virtues such as solicitude, loving kindness, and compassion, they reject that Buddhist action toward plants need necessarily be violent. Interpreting nonviolence as a virtue (rather than simply as an ethical rule) they assert:

> It would be a mistake to suppose that the virtue of non-violence expresses itself only in our relations with other sentient beings . . . it would surely be odd to suppose that someone might be gentle and considerate in his dealings with sentient beings, carefully avoiding causing them harm, and yet brutal and vicious in his dealings with non sentient beings.[131]

While at first glance this appears to be a reasonable argument, in terms of plant life, there are a number of problems with relying entirely on a virtue ethic to generate appropriate behavior toward them. Firstly, it necessitates that the virtue has been fully realized in a person's character. Therefore, in reality it can only apply to the most advanced practitioners—leaving many laypeople still susceptible to acting violently toward plants. Secondly, as Cooper and James admit, even in the case of an advanced practitioner, they would only ever act out a virtue "naturally and spontaneously in accord with the precepts."[132] My point here is that if plants are not included in the first precept, then even a virtuous, nonviolent person may still commit plenty of unintentional and often unnecessary harm to plants without even realizing.[133] As such, a virtue of nonviolence is no guarantee.

Furthermore, relying on a virtue ethic focussed on sentient beings to generate appropriate behavior toward nonsentient plants maintains the exclusion of plants from proper moral recognition. Relying wholly on this virtue ethic does nothing to change the exclusion upon which the whole Buddhist outlook on the wider natural world and, indeed life, is effectively based. As Ram-Prasad states, maintaining the exclusion of others, "is, of course . . . an act of intellectual violence; and it is the attitude that drives collective and systematic physical violence."[134] Therefore, for Buddhism as a whole to be properly nonviolent toward plants, it needs to aim for inclusion and recognition of plants as sentient beings. Certainly it requires more than a reflected virtue ethic to achieve this because of the deep-seated nature of exclusion. In excluding others from recognition and engagement:

> The Other becomes a negative necessity, that which must be set apart and kept apart for one's own sense of (collective) self to be sustained. The need for separation is vital to the continued viability of one's identity; therefore, there is a willingness to preserve the separation at high cost—to the Other.[135]

Liberating plants from exclusion, recognizing them as sentient beings, and moving toward engagement with them would constitute true nonviolent action toward the plant kingdom. Achieving this would allow plants to be recipients of solicitude, loving kindness, and compassion.

Background Reversal and Intention in Mahayana

Although there is this strong backgrounding of plants pervading the Pali and Tibetan canons, it would be misleading to portray this as the *definitive* Buddhist tradition toward plants. Buddhism is an interesting religion to examine precisely because it is extremely diverse. One of the most interesting things about Buddhist thought is that certain schools hold very different ideas about plants than those represented by the Tibetan Bhavachakra.

The emergence of very different plant perceptions in Buddhism arose with the spread of Mahayana Buddhism into East Asia during the first and second century CE. After Buddhism became embedded in China, scholars began to regard the traditional distinctions between sentient and nonsentient beings as inappropriate concepts for the portrayal of the natural world.[136] The work of William LaFleur in particular has shown that East Asian Buddhists did not follow the Indian tradition in recognizing an existential split between sentient (considerable) and nonsentient (nonconsiderable) beings. Instead they attempted to break down this divide between plants and animals by intentionally emphasizing teachings that suggest continuity and connectivity between the parts of the natural world.

The *Lotus Sutra* is a particularly influential text in this regard.[137] It contains a whole chapter dedicated to plants, which commences by noting the pervasiveness of the water which falls from rain clouds onto the Earth. This water is noted as having one essence, and it enables the trees, shrubs, herbs, and grasses that actively consume the water to grow and blossom, producing fruits and flowers. The sutra compares the rain clouds and the water to the Buddha and his teaching, which pervades everywhere and falls equally on all beings. In this analogy, regardless of their outward form, all beings are connected in a basic sense of shared substance and basic ontology. Drenched by the rain, all beings are touched equally by the teaching of the Buddha.

With an awareness of the *Lotus Sutra* teachings, Chinese Buddhist scholars in particular first began to emphasize the connections between plants and humans in their discussions of the problem of whether plants and trees could attain enlightenment. The first explicit developments of such concepts are found in the writings of Chi-t'sang (549–623 CE), a Chinese master of the Madhyamika School. He used the idea that as plants are very much like other (sentient) beings in their actions, in theory there can be an "Attainment of Buddhahood by Plants and Trees."[138] In contrast to the Indian-Tibetan tradition, which attempted to distance plants from other living beings—by concentrating on their similarities to sentient beings, Chi-t'sang pursued the idea that they had the same nature and potential for action.

This opening of the sphere of beings worthy of consideration was also attempted by Chan-Jan (711–782 CE), of the Chinese T'ien-t'ai, a school of Buddhism that regards the *Lotus Sutra* as its primary scripture.[139] Chan Jan also dissolved some of the distinction between sentient and nonsentient by professing that "even nonsentient beings possess the Buddha nature."[140] Chan-Jan did away with a ranking based upon the attribution of sentience and consciousness, and instead, he directly expounded the teaching in the *Lotus Sutra* that an awakened or Buddha nature pervades and connects *all* beings. Therefore, in this case, all beings (whether sentient or not) are worthy of consideration, and Chan-Jan asserts:

> In the great assembly of the Lotus, all are present—without divisions. Grass, trees, the soil on which they grow, all have the same kinds of atoms. Some are barely in motion while others make haste along the Path, but they will all in time reach the precious land of Nirvana.[141]

Taking the argument to its logical conclusion, if all beings possess the Buddha or an *awakened* state as their inner nature, all beings must be on the path to enlightenment. Though human perception may find a lack of speed and movement in plants and rocks, Chan-Jan put forward the idea that plants act in the same way as other volitional beings and purposely strive for enlightenment.

The teachings on the "Buddha nature of trees and rocks" were first transferred to Japan from the T'ien-t'ai tradition, by Saichō (767–822 CE), the founder of Tendai Buddhism.[142] This transition extended the general metaphysical concern of pressing Buddhist universalism to its logical maximum and including all phenomena within the realm of the Buddha nature. In the writings of the Japanese Buddhist scholars this insistence on Buddhist universalism became more narrowly focussed on the inclusion of the natural world.[143] Callicott regards this emphasis as possibly an influence of the Indigenous Shinto religion, which attributes elements of divinity to much of the natural world.[144] Parkes regards Shinto philosophy as the reason why it was accepted by all the main Japanese sects that Buddha nature is in all beings.[145] Certainly the notion that "the material never exists without *some* relation to the spiritual"[146] has strong parallels with idea that entities such as grasses, trees, rocks, and mountains are inherently enlightened (*hongaku-shiso*).[147]

Kūkai (774–835 CE), the founder of Shingon Buddhism, attempted an explicit dissolution of the boundary between the sentient and nonsentient. He did this by stating that everything in existence must have the Buddha nature as its ontological basis.[148] Therefore, for Kūkai, the very essence of vegetation was this awakened potential, or Buddha nature, and he makes this clear by stating:

> If plants and trees were devoid of Buddhahood
> Waves would then be without humidity.[149]

The explicit statement that the same awakened nature permeates all things dissolves the separation between sentient and nonsentient beings. In a plant context, according to LaFleur, Kūkai's view was that "plants and trees [have]

Buddha-nature simply because they, along with everything else in the phenomenal world, are ontologically one with the Absolute, the dharmakaya."[150]

Ryōgen's (912–985 CE) "Account of (How) Plants and Trees Desire Enlightenment, Discipline themselves and Attain Buddhahood" draws directly upon the teachings in the *Lotus Sutra*.[151] From the *Lotus Sutra*, Ryōgen infers that the life cycles of plants move through four stages, which correspond to the stages on the path to enlightenment, and so:

> Grasses and trees already have four phases, namely that of sprouting out, that of residing [and growing], that of changing [and reproducing] and that of dying. That is to say, this is the way in which plants first aspire for the goal, undergo disciplines, reach enlightenment and enter into extinction. We must, therefore, regard these [plants] as belonging to the classification of sentient beings. Therefore when plants aspire and discipline themselves, sentient beings are doing so. When sentient beings aspire and undergo austerities, plants are aspiring and disciplining themselves.[152]

Maintaining the distinction between sentient and nonsentient, Ryōgen argues for the recognition of plants as sentient beings. He makes the teachings of the *Lotus Sutra* explicit and directly connects the life cycle of the plant to the process of enlightenment experienced by a sentient being, particularly the human. He takes the familiar changes in the life of a plant and sees in them the four stage process in Buddhism of arising, continuing, changing, and ceasing to be.[153]

In contrast to earlier arguments, Ryōgen did not regard the stillness of plants as a reason for classifying them as inanimate. The work of LaFleur makes it clear that Ryōgen regarded stillness in plants in the same way as the meditative stillness found in a Buddhist monk. Possessing such a positive characteristic, in some instances, plants were considered as spiritually refined. Indeed Ryōgen saw "no better Buddhist yogis in the world than the plants and trees in his own garden, still, silent, serene beings disciplining themselves towards nirvana."[154] Such qualities of stillness, silence, and equanimity are also cherished in the early Pali canon. In this context, Findly has proposed that as they are generally stationary, plants may be seen as spiritually advanced. They do not generate, but only consume karma in this life in the same manner as a Buddha.[155]

Chujin (1065–1138 CE), a Tendai scholar, built upon Ryōgen's argument, but also took it a step further by eliminating the implicit suggestion that plants are sentient beings because their activity resembles the human.[156] Instead, for Chujin, their mere existence as living beings is enough to recognize that they too are progressing on the path toward actualization. To this effect, in his *Kanko Ruiju*, Chujin writes:

> As for trees and plants there is no need for them to have or to show the thirty two marks (of Buddhahood); in their present form—that is by having roots, stems, branches, and leaves, each in its own way has Buddhahood.¹⁵⁷

In LaFleur's analysis, Chujin "does not follow Ryōgen in forcing the members of the plant world into a frame of reference based on human experience" but instead he allows these "members of the natural world to have their own enlightenment in their own way and on their own terms."¹⁵⁸

To realize and connect with this nature in plants, it is necessary to follow the advice of the Zen poet Bashō who advised, "go to the bamboo if you want to learn about the bamboo" and "leave your subjective preoccupation with yourself."¹⁵⁹ Such wisdom could provide a suitable guide for any Buddhist attempt to seek affinity with the plant kingdom. Bashō's other caution that if you approach the plant kingdom from a human perspective, "you impose yourself on the object and do not learn" appears almost phenomenological and resonates strongly with our search for a more philosophical botany.¹⁶⁰

Although this section highlights that East Asian Buddhists have sought to recognize sentience and enlightened potential in the plant kingdom, it is important to question their motives. While scholars have intentionally recognized sentience in the plant kingdom, the question remains; did they do so in order that plants be included within the realm of moral consideration? That is, can we substantiate the notion that because plants are regarded as sentient or even enlightened—that they are automatically granted some sort of intrinsic value and are, therefore, worthy of protection? ¹⁶¹

Unfortunately, the evidence is such that it is difficult to claim this. Within the Zen tradition plants are still generally regarded as nonsentient in a moral context. As James explains:

> Zen masters . . . do not take such extreme measures to avoid impinging on the world, so it is unlikely that in claiming to have Buddha nature, they mean to accord them [plants] moral standing in a way that brings them into the purview of the first precept.¹⁶²

Such an interpretation is in line with the explicit stance of Chujin, who although rigorously maintained that plants also had the Buddha nature, made it clear that this was from the perspective of ultimate truth. From this perspective, there is no distinction between plants, animals, and human beings, but "from this standpoint moral distinctions are not relevant either."¹⁶³ As James notes, Schmithausen also agrees with the interpretation that East Asian claims for sentience, and Buddha nature in plants do not affect moral consideration.¹⁶⁴ Any attempt to construct an East Asian environmental ethic needs to address this continued position of exclusion. Within a wider environmental debate, it is a potentially significant factor in the anthropogenic conversion of much of the East Asian landscape.[165]

To counter such philosophies of plant exclusion, the following chapter will introduce more evidence for the incorporation of plants within human moral behavior. Specifically, it will consider the nature of human-plant interactions in Indigenous cultures, which explicitly seek to connect with the other living beings in this (rather than the "next") world. Whereas this case study posits ontological connectivity upon notions of karma, rebirth, and Buddha nature, Indigenous cultures offer more Earth-based principles for connecting with the other beings that we live amongst.

5

INDIGENOUS ANIMISMS, PLANT PERSONS, AND RESPECTFUL ACTION

> Listen carefully this, you can hear me.
> I'm telling you because earth just like mother
> and father or brother of you.
> That tree same thing.[1]
> —Bill Neidjie

Within wider discussions of environmental ethics, Indigenous philosophies and worldviews have been put forward as philosophical counterexamples to the Western hyperseparation from nature.[2] A summary of a general Indigenous position is provided by Whitt et al. who consider that "Indigenous responsibilities to and for the natural world are based on an understanding of the relatedness, or affiliation of the human and non-human worlds."[3] In general it can be argued that this relational approach stands in contrast to Western treatments of the natural world as a radically different, inferior Other. In the context of the present study Indigenous lifeways may "offer insights that may help dominant societies unlearn some things and become open to other ways of knowing the world."[4] In knowledge of a diversity of living Indigenous traditions and a multitude of Indigenous cultural practices, this chapter will maintain the established theme and will focus on the occurrence of specific ideas and practices related to plants. In particular, it aims to uncover how plants are incorporated within the realm of human moral considerability. It will pay specific attention to the principles that lead to respectful relationships coexisting with human-plant predatory relationships. Here I am not seeking to universalize or appropriate Indigenous thought for use in Western societies. This chapter is written "not for imitation or for direct instruction, but for inspiration and enriched understanding."[5]

This chapter will deal almost exclusively with the well-documented Indigenous cultures of Australia, North America, and Aotearoa/New Zealand. While acknowledging their distinctness, these cultures have been assembled together in their recognition as animistic societies, scholars of the "new animism."[6] Recent scholarship on animistic Indigenous societies presents a number of shared relational principles and criteria, which this chapter presents thematically.

Primarily, in animistic worldviews, it is a general principle that the plant, animal, and human realms interpenetrate. Within a great diversity of oral traditions, in almost all cases, there is a recognition of the kinship between human beings and the natural world, a kinship that is based not upon rebirth (as in Chapter 4) but upon shared heritage and substance. This idea of a shared genealogy is prevalent in the wonderful Earth ancestry stories held by Aboriginal Australians, Native North Americans, and Maori people among many others. Many of these creation stories involve the strong motif of metamorphosis to express interpenetration. These Earth-based stories are a key feature of Indigenous traditions. In New Zealand, as in other regions "Creation accounts are the foundations upon which Maori of the Pacific have built a cosmological, religious philosophy and metaphysics."[7] Due to the importance of these accounts, this chapter is structured around their narrative presence.

Less well-known than this basic Earth relatedness or the widely discussed idea of sacred landscapes is the concept of *personhood*, championed by the scholars of animism such as Irving Hallowell, Nurit Bird-David, and Graham Harvey. Rather than focus on sacred plant species, this study of human-plant interactions will be supported by the concept of personhood.[8] Personhood is a crucial, all pervading concept—for as persons, plants are recognized as volitional, intelligent, relational, perceptive, and communicative beings. Living in a world underpinned by the plant kingdom, the existence of plant persons is incredibly important for discussions of interspecies ethics.

This acknowledgement of plants as persons is based on and in turn strengthens the recognition of plants as kin.[9] Indeed, personhood is expressed and galvanised within specific kinship relationships between individual plants and humans. These specific, local kinship relationships are accompanied by obligations of responsibility, solidarity, and care. Therefore, they are one of the most important aspects of inclusive human-plant relationships. Crucially, however, in animist cultures, the recognition and acceptance of plant personhood and specific kinship coexists with predatory relationships.

In contrast to Western philosophies, Indigenous animist societies do not seek to deny personhood and pursue exclusion because some persons must be killed for others to live. In contrast to Jainism, animist lifeways do not avoid violent contact with individual plants. While they aim toward connection and inclusion, the examples of animist thought and practice included in this chapter do not lead to a totally hands off approach to the plant kingdom. The need to kill persons is accepted as a fact of life. This chapter examines the outcomes of the interplay and negotiation between the recognition of kinship, personhood, and relationships of use that involve killing.

Human and Plants as Earth-born Kin

In her elucidation of the general worldview of Native North Americans, Carol Lee Sanchez regards the "principle of relatedness" to be fundamental to the majority of Indigenous societies.[10] In Australia, Deborah Rose describes an Indigenous philosophical ecology that is based upon recognition of "pattern, benefit, and connection."[11] A basic sense of kinship between human beings and other beings of the Earth arises from the stories of the Dreaming beings—the creative beings or "spirit ancestors" that shaped the Australian landscape, giving it its present form.[12] All the Dreaming beings originated from within the earth. They walked their separate paths and generated the sets of relationships, and the rules of relationship, which are known to Aboriginal people as Law.[13] As Hobbles Danyari says:

> Everything come up out of ground—language, people, emu, kangaroo, grass. That's Law.[14]

Here the interpenetration of human and plant takes place in this life. This is an immediate, earthly kinship. The Dreaming beings emerged from the ground and are linked in basic kinship—in Law, by being earth-born. Such recognition of ancestral relatedness is coupled with a direct recognition that each being is also autonomous in this life. Each Dreaming being walks the Earth independently and "every life form that came out of the earth, and every modern descendant of an original life form is still autonomous."[15]

Although many Dreaming stories concern animal species, a significant number also involve plants. The Adnyamathanha of the Flinders Ranges in Southern Australia tell the story of a metamorphosis concerning the *Iga* tree (*Capparis mitchelii* Lindl.). The *Iga* was initially a Dreaming being in the shape of a man who had travelled south from Queensland to find a woman in the country (ancestral homeland) of the Adyamathanha. In the act of digging for *ngarndi* roots, the man and woman were scared by something, and they said:

> Well, look, if we turn into a tree, they mightn't take any notice of us. So they came along as trees from there.[16]

The *Igas* transformed from their initial humanity into the form of those trees that continue in existence today. The transformation of these Dreaming beings created the persisting lineage of the *Iga* and also the lineage of the Adnyamathanha people. A similar story is told by the Ngulugwongga of the Daly River in northern Australia, of how the red lily came to be in their country. The Dreaming Yilig-moi-indih or "Red Lily Woman" came from another country carrying lily roots, which she planted in the ground. After she had finished her planting, Yilig sat down on the ground. Along with all the other lilies she went down into the ground, and that is where she is today.[17]

Not all plants in the Dreaming stories have an initial humanity, but all are autonomous and can trace their origins to the actions of a Dreaming being. In the Ngarinman account of creation, it is clear that plants have their own Law; the "Plants started growing according to their own 'laws'—their own shape, size, habitat requirements, and "behaviour."[18] These beings of the earth are subject to their own Law, but are distributed throughout country by the actions of Dreaming beings. For Yanyuwa people, the cycad palms grow in their country as a result of the Tiger Shark who "threw the cycad nut everywhere, over long distances he threw it."[19] This traditionally important food for the Yanyuwa has its origins in the actions of the spirit ancestor.

Thus, the Dreaming beings are not just the spirit ancestors of human beings, they are ancestral to humans, animals, and plants and "from them it is possible, and indeed imperative, to trace kinship among the things of the world."[20] The Dreaming stories provide basic ancestral genealogies that "for Indigenous peoples typically do not confine themselves to the human."[21] Such stories have their parallels in the *Song of Purusa* from the *Rig Veda*, in which all beings share origins, but in Aboriginal Australia, they are more local. These ancestral connections provide the basis of a living kinship and an ethic of connection between all beings in the natural world—beings that are ultimately derived from the Earth.[22] Dreaming beings are the ancestors of every constituent of a sentient landscape.[23] For Gagudju elder Bill Neidjie, whose home country is in Kakadu National Park, North Australia, this kinship is an intimate experience:

> That tree, grass . . . that all like our father.
> Dirt, earth, I sleep with this earth.
> Grass . . . just like your brother.
> In my blood in my arm this grass.[24]

Such basic, consubstantial kinship is characteristic of Indigenous worldviews. As well as in Australia, kinship is also prominent in the creation stories of Indigenous North Americans. Oglala speak of the "Sacred Hoop" of the cosmos, within which everything is bound by *wakan* (sacred power). According to Fritz

Detwiler, "as part of the Hoop, all beings are related in a way that reflects the ontological oneness of creation."[25] The stories of the Oglala Lakota describe a network of kinship and ancestry that links all beings in the natural world. Detwiler quotes Wallace Black Elk:

> Our real Father is Tunkashila [Creator], and our real Mother is the Earth. They give birth and life to all the living, so we know we're all interrelated. We all have the same Father and Mother.[26]

The Oglala Lakota phrase *mitakuye oyasin* ("all my relatives") expresses this sense of kinship and ancestry that links people with plants, animals, rocks, and waters.[27] The recognition of relatives extends into ritual and ceremony. Kenneth Morrison writes that "Native American prayer acts are commonly invocations of kinship, at once earnest petitions and reminders of interdependence."[28]

Maori people also have a creation story, which unites plants and people in genealogical descent (*whakapapa*). In the original Maori homeland, Hawaiki, the Sky Father (Ranginui) and the Earth Mother (Papatuanuku) were "lovers locked in an age-long embrace, during which they had many children."[29] The children of these two gods were the progenitors of the entire natural world—including plants, rocks, seas, winds, animals and human beings. In Maori philosophy, all the beings of the natural world share *whakapapa* from Papatuanuku and Ranginui, and also from Tane-mahuta, whose actions created the world. Erenora Pukatepu-Hetet explains that the flax plant (*harakeke*) "is a descendant of the great god Tane-mahuta . . . today's Maori are related to *harakeke* and all the other plants, Tane is their common ancestor."[30] All beings share a basic common ancestry and knowledge of a being's *whakapapa* is fundamental to Maori epistemology. Roberts and Willis explain that:

> To "know" oneself is to know one's *whakapapa*. To "know" about a tree, a rock, the wind, or the fishes in the sea—is to know their *whakapapa*.[31]

It is vitally important that all beings have a shared *whakapapa* so that all beings can be located in the world. As is the case in Aboriginal Australia, "Things that are deemed to have no connections, which are not related to anything at all, can at their worst be utterly meaningless."[32] Human beings are embedded within the natural world partly because they have a worldview that recognizes shared ancestry with other beings. This worldview is "at the heart of Maori culture, touching, interacting with and strongly influencing every aspect of it."[33] Therefore:

> Maori people do not see themselves as separate from nature, humanity and the natural world being, direct descendants of the Earth Mother.

Thus, the resources of the earth do not belong to humankind; rather humans belong to the earth.[34]

As an integral part of nature, Maori have responsibilities of care toward other species as well as other human beings.[35]

As hinted at in the Aboriginal Dreaming stories, as well as ancestry, another recurring motif is one of shared substance. This is intimately linked with notions of genealogical descent and is commonly expressed through the theme of transformation. For the Koyukon of Alaska, all things were created in a time known as *Kk'adonts'idnee,* which Richard Nelson translates as Distant Time.[36] Nelson writes:

> During this age "the animals were human"—that is, they had human form, they lived in human society, and they spoke human (Koyukon) language. At some point in the Distant Time, certain humans died and were transformed into animal or plant beings, the species that inhabit Koyukon country today.[37]

There is shared ancestry, but here also predominantly a metamorphosis of substance between human beings, plants, and animals.[38] At the end of Distant Time, a great flood covered the Earth and while some plants and animals survived, "they could no longer behave like people."[39] All human beings were killed, and Raven had to make them anew, in the form they are today. The neighboring Tlingit people say that Raven reconstituted these new people out of wood and leaves, which is why the human is now a mortal.[40] Humans share mortality with the trees from which they were crafted by Raven.

A similar plant origin is told by the Tsimshian, who say that the elder bush gave birth to human beings—bestowing on us their soft skin and, crucially, their mortality.[41] In addition to shared birth, another motif of shared substance is shared possession of blood, which for a plant is its water and sap.[42] For Bill Neidjie, this unites humans and plants as kin sharing the same substance:

> That's your bone,
> your blood
> It's in this earth,
> same as for tree.[43]

Plant Persons

Recent revisitations of animist cultures have rejected the "old animism" of Tylor et al.—that of a failed epistemology based on a naïve, misguided belief in nature spirits.[44] Instead, academics of the "new animism" such as Nurit Bird-David and Graham Harvey have begun to recognize human interactions with this earthly world in terms of kinship links and relationships between persons.[45] Graham Harvey describes animists as "people who recognise that the world is full of persons, only some of whom are human, and that life is always lived in relationship with others."[46] Rather than a deluded belief that everything is alive, animism is a sophisticated way of both being in the world and of knowing the world; it is a relational epistemology and a relational ontology.[47]

This recognition of personhood is one of the most important elements of the "new animism" and is of great significance for this study. This is because personhood is recognized in animist societies which are also famous in an ethnobotanical context for their extensive use of the natural world. It is clear that acknowledgement of personhood and the use of plants are not incompatible.

The combination of relatedness and autonomy is fundamental to an understanding of an animist worldview. Relatedness and autonomy are recognized in the application of the term *person* to beings and phenomena that stretch way beyond the human. A fundamentally shared ontology necessitates a heterarchical continuum of persons rather than the construction of ontological hierarchies. The use of the term *person* is drawn from Irving Hallowell's seminal work with the Ojibwa of North America. In his analysis of the relationships between human Ojibwa and the characters of mythology and story, Hallowell recognized that "to the Ojibwa they are living "persons" of an other-than-human class."[48] For the Ojibwa, animal, plant, stone, and sky beings are also considered to be persons.[49]

Phillipe Descola stresses that talking of persons who are not humans is not a journey into metaphor. Animist societies treat other beings "as proper persons."[50] Using Hallowell's concept of personhood as a basis, Graham Harvey regards the recognition of other-than-human persons as fundamental to an animist worldview. Defining what it means to be a person, Harvey explains:

> Persons are those with whom other persons interact with varying degrees of reciprocity. Persons may be spoken with. Objects by contrast, are usually spoken about. Persons are volitional, relational, cultural and social beings. They demonstrate agency and autonomy with varying degrees of autonomy and freedom.[51]

It is important to be clear that this recognition of plant personhood is not anthropomorphic.[52] A worldview that relates to other-than-humans as persons is not concerned with projecting human-like qualities where they do not exist. Nor is it a case of "confusion between persons and objects."[53] As Harvey states, "To be a person is to want to continue living."[54] As persons, plants are not naively thought to have human faculties. They are understood to be living beings with their own perspective, and with the ability to communicate in their own way. Personhood thus emerges from a focus on relating and the recognition of shared volition and intentionality in natural beings. As such, although it is supported by the transmission of ancestral knowledge and wisdom, recognition of personhood is not genetically determined; it must be learned.[55] The recognition of personhood (and importantly how to act toward persons) "is found more easily among elders who have thought about it than among children who still need to be taught how to do it."[56] In this way, it is a personal, dynamic way of coming to know and encounter the world.[57]

While the term *person* is drawn from the Ojibwa, the recognition of personhood can also be located in other cultures. For the Oglala, any of the beings within the "Sacred Hoop" "are power centers, intentional manifestations of power who act according to their particular character."[58] That is, as Detwiler emphasizes, all beings in the Oglala world view are considered to be persons:

> The Oglala understand that all beings and spirits are persons in the fullest sense of that term, they share inherent worth, integrity, sentience, conscience, power, will, voice, and especially the ability to enter into relationships. Humans, or "two-leggeds" are only one type of person. Humans share their world with Wakan and non-human persons, including human persons, stone persons, four-legged persons, winged-persons, crawling-persons, standing-persons (plants and trees), fish-persons, among others. These persons have both ontological and moral significance. The category person applies to anything that has being, and who is therefore capable of relating.[59]

From a shared ontological ground of existence, this person-oriented worldview is shared by many of the Native American tribes of the Pacific Northwest coast.[60]

For Australian Aboriginal people, persons are embedded within a local landscape, or "nourishing terrain" that is itself a subject.[61] In Australia, the land, or more commonly "country" is "seen to have a will, a life force of its own . . . it can know individuals or be ignorant towards them."[62] Deborah Rose describes country as a place that "gives and receives life" and notes that country is a living entity with its own volition and consciousness.[63] Within these sentient landscapes, personhood is present in all the constituents and participants in country. Rose explains:

> Subjectivity, in the form of consciousness, agency, morality and law is part of all forms and sites of life, of non human species of plants and animals, of powerful beings such as Rainbow snakes, and of creation sites, including trees, hills and waterholes.[64]

With this awareness, country can be understood as a person, itself composed of a multitude of unique, individual persons. Thus, for Aboriginal Australians, country is not a hierarchy, but a heterarchy of related beings. In Aboriginal understanding, all persons in country are saturated with sentience because all can trace their origins to the actions of the Dreaming ancestors. In this context, Rose states:

> For many Aboriginal people, everything in the world is alive, animals, trees, rains, sun, moon, some rocks and hills, and all people are conscious. . . . All have a right to exist, all have their own places of belonging, all have their own Law and culture.[65]

Personhood stems from kinship and shared ancestry, which ultimately arise from the Dreaming—not just a time past, but the ever present "everywhen."[66] Thus, the immanent presence of the Dreaming imparts autonomy to all beings. Bob Randall, an Elder from Central Australia, says "not only is the natural environment a source of food, it is also the expression of *tjukurrpa,* the Dreaming."[67] As the Dreamings instituted the lineages of each species, each natural being is recognized to be operating under its own laws and culture.

For certain plant lineages, Dreaming stories tell of their initial humanity and then their metamorphosis into the beings that exist in-country today. This transformative process is an expression of the existence of personhood as well as kinship. The attribution of initial humanity to plant species is not anthropomorphism but a significant way of "representing interiority" or subjectivity.[68] This is no "primitive" projection of our own humanity, but a complex system "for encountering the world."[69]

To provide an example, one Gunwinggu Dreaming story relates that the pandanus trees are the transformed couple, Namalbi and (his wife) Ngalmadbi who left their camp after a quarrel with their family.[70] In another Gunwinggu tale, Mananda was an old man from South Goulbourn Island who was unable to walk very far. One day he told his sons that he would remain in one spot while they went off and "He just sat there for so long that he became a long yam."[71]

Stories of transformation and metamorphosis recognize plants (and other beings) as volitional, communicative subjects. This recognition in turn provides the basis for dialogical relationships between plant persons and other persons.[72] Presentations of plants as persons even occur in stories where an initial humanity is not explicit. In the story of the *Igas* in Adnyamathanha country, it is only the *Igas* that are described as once being human, but the other trees also display many of the attributes of personhood. In the story, the mulgas and the coolabahs, demonstrate their personhood (and territoriality) by attacking the *Igas*.[73]

As well as transformation, personhood is manifested in communication, one of the necessities of relating. In this context, Aboriginal elder Bill Neidjie succinctly describes the awareness and communicative capacity of the plant persons in Gagudju country:

> Tree. . . .
> he watching you.
> You look at tree,
> he listen to you.
> He got no finger,
> he can't speak.
> But that leaf. . . .
> he pumping, growing,
> growing in the night.[74]

In Neidjie's description of plant communication, it is apparent that the tree is aware of the human being, and expresses its own "voice" by growing. This dialogical flourishing is a theme that is taken up in the final chapter as a model for human moral consideration of plants.

In Northwestern North America, the Bella Coola have stories of a time when human beings and trees could talk to each other. Although the common language has gone, it is said trees can still understand human speech, which means trees can be spoken to and addressed with prayer.[75] Plants may not be able to talk in human language, but they are aware and they have their own methods of communication.

Back in Australia, in Yanyuwa country, human beings communicate directly to plants through human language. People address songs directly to the cycads in Yanyuwa country in order to keep the trees healthy and to ask them to produce fruit.[76] Indeed, some of the best examples of plant-human communication are from Australia. Across the continent, Aboriginal country is spoken with. Aboriginal people recognize that communication flows through country "between individuals, groups and species."[77] In the Kimberley region of Western Australia, Paddy Roe and Frans Hoogland stress the need to cultivate a feeling for country in order to perceive this communication between its elements, the animals, the rocks, and the trees.[78] Bill Neidjie also makes clear that through being aware and open toward country, you can be aware of the communicative possibilities in the trees:

> Feeling make you,
> Out there in open space.
> He coming through your body.

> Look while he blow and feel with your body. . . .
> because tree just about your brother or father. . . .
> and tree is watching you.[79]

The reciprocal communication from plants can manifest in a variety of ways. As Neidjie points out, all these ways are linked to plant vitality and growth. Aboriginal people often know that plants can tell you about other things that are happening in country. In Wik-Mungkan country, when the *thanchal* tree flowers [*Alstonia actinophylla* (A.Cunn.) K.Schum.] it indicates that the oysters are fat enough to eat.[80] For the Tiwi of Bathurst and Melville Islands, the flowering of the tall wet season grasses such as *Sorghum plumosum* (R.Br.) P.Beauv. tell of the end of the wet season. They also communicate the arrival of migrating birds such as the black-faced cuckoo shrike.[81] At Daguragu, when the *jangarla* tree flowers, [*Sesbania formosa* (F.Muell.) N.Burb.], it relates that the crocodiles are laying their eggs.[82] All these instances are flows of ecological communication from the plants in question to human beings and beyond, to the species they concern.[83] Awareness of the communicative capacities of these plant species requires openness, awareness, and knowledge of country and plants.

Relationships of Sameness and Difference

In the Indigenous cultures that I have briefly touched on, plants are persons, but they are not naively regarded as being identical to human beings. Plants, animals, and humans are acknowledged as possessing different attributes. Some Native American Indigenous traditions can mark the difference between human, plant, and animal in terms of varying amounts of power. As Irving Hallowell says of the Ojibwa worldview:

> In relation to myself, other "persons" vary in power. Many of them have more power than I have, but some have less.[84]

Unlike in certain Western modes of thought, this recognition of different powers is not a value-ordered hierarchy, which seeks to represent humans as superior and other beings as inferior. For different language groups, plants occupy various positions on a continuum of expressed power. According to Richard Nelson, for the Koyukon, some (but not all) plants have "spirits," "vaguely conceptualised essences that protect the welfare of their material counterparts."[85] The spirits of the plants are generally less powerful than those of animals and are not as vengeful toward wasteful and irreverent behaviors. The human being is markedly different in that:

> Only the human possesses a soul . . . which people say is different from the animals' spirits. I never understood the differences, except that the human soul seems less vengeful and it alone enjoys immortality in a special place after death.[86]

Perhaps the use of the terms *spirits* and *soul* are inappropriate in light of the reconsideration of animist cultures in terms of personhood. Indeed, Richard Nelson acknowledges that the spirit and the material being are one and the same thing.[87] Interestingly, the Oglala appear to invert this schema of personhood. Although all beings have a shared ontology, there are three categories of persons based upon varying degrees of power. The first category is the *Wakan*, including beings such as the Earth, Sun and Sky, which possess the greatest power.[88] The second category is the nonhuman, including plants and animals. The plants and animals are regarded as more powerful than human beings. Detwiler describes how nonhuman persons and human persons differ:

> These non-human persons have sentience, will, and voice to a greater degree than human beings. And, unlike humans, non-human persons engage more easily in harmonious kinship relations with other persons in the Sacred Hoop. When that harmony is disrupted, non-human persons respond in kind to those who fail to treat them with respect and beneficence. When treated respectfully, they respond generously. Their behavior may be seen as beneficial or as harming depending upon human need and ethical awareness. Non-human persons typically communicate with human beings through dreams and visions. They often indicate their willingness to establish kinship relations with humans. Such relations involve learning the song that is the non-human person's mode of communication. The song gives humans access to the non-human person and makes them more aware of the relational ethics which constitute the Sacred Hoop.[89]

Unlike the Koyukon, the Oglala regard animal and plant persons as having greater power as they are better at relating to the world around them. The emphasis and importance of relationality is laid bare. Human beings who are able to communicate with these nonhumans through song and ceremony are able to tap into the greater power of these persons.[90] Those humans that neglect such necessities are regarded as the most destructive type of person, "as a result of failing to harmonize their will with other-than-human persons."[91] Oglala ontology also places humans at the bottom of a power–ranking because of human (total) dependence on other-than-human persons for our existence. This recognition fosters humility and moral sensitivity in humans, thus reinforcing the requirement for mutual relations between human and other-than-human persons. The lack of such humility and the failure to harmonize the needs of all types of persons is prevalent in contemporary Western society and has relevance for the current anthropogenic environmental crisis.[92]

Here there is an obvious connection to the appropriate human response in the face of obvious alterity. In the Western streams of thought dealt with in Chapters 1 to 3, difference is construed as radical and is the basis of a drive toward separation and exclusion. In the animist societies mentioned here, the response to difference occurs within a relational framework and so is one of engagement and inclusion, Rose's "ethic of connection."[93] Indeed, for proper relationships with other persons, it is essential to recognize both similarity and difference.[94] Within animist worldviews, even though they may be different, other living persons are valued for their role in life and for their knowledge of the world that they possess.[95] In the terminology of Bakthin, it could be said that animist worldviews allow the "voices" of nonhumans to be heard, thus opening up the possibility of learning from them.[96]

Harming Plants

Although plants may be considered as persons, it is obvious that in order for human beings to live they must violate plants' integrity, curtail their flourishing, and ultimately harm them. Human persons must sometimes act against the interests of herbs, shrubs, and trees that are actively striving to live and reproduce. In animist cultures, "the aliveness or personhood of a person is no guarantee that they will not be killed as food or foe."[97] This stands in sharp contrast to the ideals of *ahimsa* for sentient beings in the Jain and Hindu texts.

In the face of alterity, Indigenous animist cultures seek connection and recognize personhood. They also partake in predatory relationships, which in turn, act as markers of difference. The need to harm and take the lives of other persons in order to live is accepted as a messy fact of life. Indeed, in many animist cultures, predatory relationships are a part of engaging with other-than-human persons; a way of acknowledging the roles that they play in underpinning human existence.[98] A key feature of Indigenous cultures is that the harm done to individual plants is not ignored or backgrounded. Significantly, there is no attempt to jettison kinship links and to undermine the qualities of plants in order that human beings might pretend that their lives can operate without harming the integrity of other beings.

Animistic societies are acutely aware of violating the integrity and autonomy of other persons. The Maori must cut the forests that they are related to and also dig up the sweet potato tubers, which are their kin, in order to provide food and shelter for their guests.[99] In some animistic societies, people must use plants that are explicitly recognized as having the capacity to suffer (and so be harmed). Amongst the herdspeople of southern Arabia, there are stories that describe the suffering of the tree *Acacia tortilis* (Forssk.) Hayne, the use of which is vital to human survival.[100] This *Acacia* is acknowledged to have the hallmarks of personhood (including the capacity to bleed when cut), but it also underpins the existence of many of the tribes in the region. In North America, the Tsimshian and the Tlingit tribes recognized that trees have a capacity for feeling and suffering, but still put them to use in the necessities of life.[101]

The fact is that although human beings and plants are kin and both persons, the necessity of violating the autonomy of plants, and harming them, is recognized and accepted. In the understanding of T. P Tawhai, this recognition is a spur toward human religious activity "which seeks permission and offers placation."[102] It does not seek repentance, for the taking of life is an integral part of the world and a necessary part of living.[103]

Taking of life is accepted as a necessity, but this does not leave the door open for indiscriminate killing. There is no justification for an untrammelled use of plants by human beings. Crucially, killing and harm takes place within structure, balanced, and reciprocal relationships. Indigenous peoples are acutely aware that ecological life is connected in a system of flows that require violent action on the part of the human. Deborah Rose's insightful analysis of Indigenous ecologies shows that such connectivities position humans, plants, and animals in networks of mutual responsibility, where life "is for others as well as for itself."[104] In Aboriginal Australia, country *must* be used; plants and animals must be killed in order for country to remain healthy.[105] This is because country's health depends on the flourishing of the living beings who inhabit country, and living beings cannot live without nutrition exchanges.

Respectful Actions Toward Plant Persons

Within balanced and reciprocal relationships, violating the integrity of other beings and curtailing their flourishing should only be done where necessary. This is fundamental to respectful action toward other persons. Harm to plants (both individuals and species) can be lessened by using knowledge of their particular qualities, such as the ability of plants to regrow from their parts. For example, when digging up yams, Aboriginal people habitually leave part of the yam in the ground and cover it back up so that the yam will regrow. Leaving the yam uncovered causes unnecessary harm, both to the yam that will die, the yam species as a whole, and to the people who will be left without food. As Bill Neidjie succinctly says:

> You leaving hole
> you killing yam.
> You killing yourself. . . .[106]

As well as sharing an interdependent kinship, it is clear that plant, animal and human bodies also interpenetrate. This is another strong reason for only committing harm to plants where necessary. Burning the grass is not disrespectful as it does not cause fatal harm to the grass; when the rains come, the grass comes back again. However if the use of herbs, grasses, and trees involves killing, Aboriginal people who have a feeling for their country can feel the harm toward their individual kin. Although this is most often expressed in totemic relationships with shared "flesh," harmful behavior toward more general kin can also be felt:

> If you feel sore. . . .
> headache, sore body,
> that mean somebody killing tree or grass.
> You feel because your body in that tree or earth.
> Nobody can tell you,
> you got to feel it yourself.[107]

Intriguingly, this is not simply a sentiment found in Indigenous cultures. The interpenetration of plant and human is simply expressed by one of the English languages most eloquent poets, Andrew Marvell (1621–1678 CE) in his work *The Mower's Song*:

> What I do to the grass, does to my thoughts and me.[108]

Perhaps it is necessary to feel this pain in order to appreciate the life that has been taken so that human beings may live. Full awareness of the interpenetration of human and plant bodies could help avoid the worst type of killing—the overexploitation of our kin and, thus, the severing of the connectivities on which life is founded.[109] The following account of the Mohawk method of gathering plants expresses the strong awareness of the need to respect the autonomy of other species and to maintain the connections between them:

> What I was taught was that when you see a plant, to first see that it's the one you offer thanksgiving to, that plant is still here with us, still performing its duty and that you wish it to continue. You walk past it and you look for the other one, and that one you can pick. For if you take that first one, who is to know, maybe that's the last one that exists in the world.[110]

As well as for species, respect is necessary for individual plants. On the Northwest coast of North America, traditional resource utilization was geared toward respectful relations with plants. Specifically, people tried to lessen the harm done to trees by taking planks from standing trees, and by not taking so much bark from trees that it impeded sap circulation and killed the tree.[111] Before felling trees out of necessity, Kwaikiutl and Tlingit would often pray and offer gifts to the trees, while also asking for permission from them.[112] Respect was also afforded to the plants in the vicinity, and when felling trees, the Tlingit would scatter eagle down on the ground to prevent further injuries.[113]

In this treatment of the harm that humans do to plants, the final stop must be to note that in many animistic cultures, death is often not perceived as a termination, but rather a transformation from one phase of life to another. It is a salient feature of animistic ways of life:

> The transformation of living persons from trees to "artefacts" is not experienced as a destruction of life and personhood, nor their consequent transformation into artificiality. Human artefacts not only enrich the encounter between persons, but are often themselves experienced as autonomous agents.[114]

The death of a plant can be particularly transformative if it enables greater connection between other persons. For Native Americans, tobacco has many important ceremonial uses in which the plant is not only considered to be sacred, but also to be a living person with the attributes of personhood.[115] To the Kickapoo nation, tobacco manifests as a powerful person and is offered to other persons through ritual.[116] The sharing of tobacco and the sharing of its smoke deepens relationships between persons and extends communication.[117] This ability demonstrates the continuance of personhood.

However, the fact that the taking of life can be transformative does not sanction excessive consumption, waste, or negligence. The life that is taken is a gift, and "To use that which is extra to the gift is immoral, ungrateful, antisocial, greedy and insulting."[118] To the Kwaikutl of North America, any objects made of cedar wood retained the characteristics of personhood as they were made from living persons.[119] For the Maori artists that weave the flax that they are related to, respect for the life of their kin is critical. As Puketapu-Hetet relates:

> It is important to me as a weaver that I respect the *mauri* (life force) of what I am working with. Once I have taken [flax] from where it belongs, I must give another dimension to its life force so that it is still a thing of beauty.[120]

Specific Kin and Care

On one hand, there can be difference and necessary harm, but on the other, there is appreciation and inclusion. The recognition of shared ancestry provides the basis for specific relationships of reciprocal responsibility with the other creatures of the natural world. In the Indigenous cultures on which we have briefly touched, there are often specific kinship links with those other-than-human species that are found in the ancestral area.

In some cultures, intimate kinship links are actually formed with plants that are the staple foodstuff or the basic building material for human beings. Rather than backgrounding, the importance of these plants manifests in an intense appreciation of the *active* role which they play in sustaining human life. As Bill Neidjie states powerfully:

> I love it tree because e love me too.
> E watching me same as you
> tree e working with your body, my body,
> e working with us.[121]

For the Maori, the sweet potato, *kumara,* [*Ipomoea batatas* (L.) Lam.], is a traditionally staple foodstuff, and the *kumara* is regarded as intimate kin. This is partly because the *kumara* migrated with the Maori in their canoe to Aotearoa, and partly because the Maori recognize that both plant and human need each other to survive.[122]

Similarly, the people of the Northwest coast of North America recognize the central role that the cedar tree plays in the continuance of human life. The trees are often addressed with terms of affection and kinship such as Long-life Maker, Life Giver, Healing Woman, or Friend and Supernatural One.[123] In Amazonia, women of the Achuar tribe relate to the plants under their cultivation as "plant children."[124] This relationship of consanguinity is very interesting because it is not understood to arise from a common ancestry. Instead it is based upon the plants and humans sharing the same living space.[125]

Once again, Aboriginal Australia has some of the most prominent instances of specific kinship links between humans and plants. In an excellent revisitation of the phenomena of totemism in New South Wales, Rose et al. explain that one aspect of totemism is a mode of sociality that:

> Articulates a system of kinship with the natural world . . . [and] is expressive of a worldview in which kinship is a major basis for all life, in which the natural world and humans are participants in life processes.[126]

The word *totem* is derived from the Ojibwa word *ototeman*, which literally means "uterine kin."[127] Totemic relationships are above all specific relationships, defined kinship relationships with other species. Totemic relationships are "relationships of mutual care" that "constitute a major system for linking living bodies into structured relationships of sameness and difference."[128] In the recognition of kinship, there is recognition of how kin differ, but this marked boundary between species is not exclusive. It is overcome by reciprocal responsibilities of mutual care and by an emphasis on consubstantiality, connectivities, and intersubjectivity. Awareness of the ecological interdependence of all life leads to a recognition of corporeal interpenetration between different living beings. It must be pointed out that the majority of totemic relationships are with animals, not with plants. However, this need not be regarded as a rejection of kinship, but a consequence of the reality that "persons identify most intimately and associate themselves most often with those who are more like them than unlike them."[129]

The inclusion of plants in totemic relationships is of great interest as they are a powerful system for incorporating specific plants into immediate relationships of care and responsibility. People and plants are directly related, and familiar kinship terms are used for species involved in these intimate relationships. As such the maintenance of totemic relationships guards against slipping into abstract generalizations of kinship and ensures that the plant persons of particular places are incorporated into the moral sphere.

Humans can have many familial totems with which they share a genealogy. While many of these heterarchical relationships are with animals, plants also feature in totemic systems. Familiar relationships are a feature of these totemic systems that include plants. In the Mak Mak homelands, totems are called *ngirrwat*. Mak Mak women Nancy Daiyi and Kathy Deveraux explain their relationships with plants:

> This tree here, we call "uncle" this tree.[130]

> Stringybark is for the women, and woolybutt is for the men. They call it "uncle." So, we're not just related to "ngirwatt" for animals. We've got relationships to trees too. That's Mum's uncle, stringybark.[131]

In Yanyuwa country, Annie Isaac tells of the grey mangrove [*Avicennia marina* (Forssk.) Vierh.] as the clan totem for the Wuyaliya clan:

> That tree, the grey mangrove, is my most senior paternal ancestor and we people of the Wuyaliya clan name ourselves as these people who are kin to the grey mangrove.[132]

In the Yanyuwa system, plants and animals are incorporated into the kinship system by being assigned a clan. This is a result of the specific actions of the Dreaming ancestors. With each species incorporated into clans, the totem and the "kin position" it is assigned to become co-extensive. So when Minnie Wulbulinimara calls a pandanus "my mother's mother's brother" or the grey mangrove "husband," she is not speaking in metaphors or symbols, but describing actual kinship relationships.[133]

In Yarralin, the matrilineal totemic system known as *ngurlu* includes the kurrajong tree [*Brachychiton paradoxum* Schott].[134] This is a totem that comes from a person's mother, and so the mother's and child's *ngurlu* are identical. In this intimate relationship, the totem and the human are regarded as kinsfolk, and in this open system, the "relationships within and between *ngurlu* open outwards to plants, animals, the elements and seasonality."[135] Unlike for the specific kinship relationships between humans and important food crops, Aboriginal totemic relationships often entail avoidance of harm and predation. Because the "flesh" of the totem and the body of the people are regarded as coextensive, people avoid eating their totem. Instead the relationship is more centered on mutual "nurturance and care."[136] These relationships, perhaps all totemic relationships, help bridge differences between persons in-country that can arise from the need for predation on kin.

Writing on these connectivities in Indigenous ecologies, Deborah Rose emphasizes that we must seek to "recuperate connection" with the other personalities in the world, but that we must do so "without fetishizing or appropriating Indigenous people and their culture of connection."[137] While this case study uses Indigenous animist ideas to inspire more respectful engagement with plants, it does not seek to appropriate Indigenous knowledge systems, or transplant them to the Western world. In any case, it would be misleading to set up a sharp dichotomy between Indigenous and European relationships with the plant kingdom. Such dichotomies are not desirable because there is evidence that before Christianity spread across Europe, pagan peoples also had respectful relationships with the plant kingdom and the wider natural world.[138] The following chapter examines this evidence and addresses its impact on some contemporary Western relationships between humans and plants.

6

PAGANS, PLANTS, AND PERSONHOOD

> A tree talked, a fir tree sighed
> an oak skilfully answered,
> "I have worries of my own
> without worrying about your son. . . ."[1]
> —The *Kalevala*

In a plant context, one of the most common claims about paganism is that pagans held certain plants and groves as sacred.[2] In the field of environmental ethics, scholars such as Lynn White Jr have used this idea of sacred nature to suggest that pagan peoples possessed an intrinsic respect for nature and maintained checks and balances on its use, something lacking in modern society.[3]

White contrasts this pagan attitude to nature as *sacred* with an anthropocentric Christianity that perceived nature purely as a resource, a proposition which has stirred much controversy.[4] Although an anthropocentric Christianity is much debated, Chapter 3 leaves us in little doubt that the creation stories in the Bible are zoocentric. They create a fundamental schism between animal life and the rest of the plant-dominated natural world.[5] Despite this, the historian Ronald Hutton doubts that the Christian attitude to nature was any more exploitative than the pagan, pointing to the Christian sanctification of certain natural forms such as springs and the pagan felling of forests.[6] Yet it can be argued that both White and Hutton fail to distinguish the pagan perception of plant life from the Christian by basing their analyses wholly on an idea of sacredness that implies the preclusion of use. It is my contention that a more important difference between the pagan and the Christian view of plants is not to be found in ideas of sacredness but in understanding how our pagan ancestors related to plants in a more ordinary sense.

Insights from contemporary animisms help reveal that in comparison with Christian texts, pagans had very different relationships with plants and a very different understanding of plant being. In sharp contrast to the backgrounding of plants in biblical and Christian theological sources, the surviving fragments of texts from pre-Christian Europe demonstrate that pagans also recognized plants as kin and as persons. Rather than backgrounding plants, the pagan material from an array of traditions—ancient Greek, Old Norse, Anglo Saxon, Celtic, and Karelian—depicts plants as relational, volitional, and autonomous living beings.

Rather than being based purely on ideas of sacredness, it is clear that respectful pagan relationships with plants were based upon the idea of relatedness and connection.[7] As in contemporary animist societies, it appears that plants were kindred beings embedded within local relationships of care, solidarity, and responsibility. Importantly, these ancient worldviews have partly inspired the establishment of contemporary Paganisms that seek the same respectful interaction with the natural world.[8] While contemporary Pagans do not always explicitly draw on the materials that I will discuss, it is apparent that they recognize and emphasize the connections between living beings. They seek to reconstruct the kinship-based relationships between plants and people that were evident in pre-Christian times. One of the aims of this chapter is to highlight the fact that this contemporary drive toward connection and respect is supported by a number of pagan sources.[9]

Pagans' Sacred Plants and Plant Kin

The sacred plants and sacred groves of pagan Europe have been written about since antiquity. Pliny relates that the mistletoe and the oak on which it grows were both sacred to the Druids and wrote that the Druids "perform no sacred rites without oak leaves."[10] Similarly, Lucan described the Druid preference for these rites to be conducted within groves of trees, perhaps of oaks.[11] This is corroborated by Tacitus who described the existence of the sacred groves of the Druids of Anglesey, which were destroyed by Suetonius Paulinus.[12] In his *Germania*, Tacitus also wrote that the German peoples "dedicate groves and woods and call by the name of gods that invisible thing which they see only with the eye of faith."[13]

Across Europe, the ubiquity of sacred groves is exemplified by the incorporation of the word *nemeton* (sacred grove) into place names across the continent. In Britain, these include *Vernemeton* in England and *Medionemeton* in Scotland.[14] In Ireland, the old name for sacred tree *bile* is too found in the place name *biliomagus* (the plain of the sacred tree).[15] In the Pyrenees area of France, altars have been found dedicated to "Fagus" and to "God six trees," and alters have been found decorated with trees.[16] Similarly in Germany, sacred spaces such as offering pits have been found to have been planted with tree trunks or living trees.[17] In Southern Europe, there is often reference to the worship of gods within sacred groves. The worship of Zeus at Dodona is described as being originally performed within a sacred grove comprised of oaks.[18] In Rome, Pliny records that the oak was a sacred tree of Jupiter, the laurel sacred to Apollo, the olive to Minerva, the myrtle to Venus, and the poplar to Hercules.[19] Ovid also reports the oak as sacred to Jupiter.[20] It seems that the oak was particularly sacred in Europe—being associated with Zeus in Greece, and with an equivalent thunder god in Celtic Europe.[21]

As the above examples show, it is common in writings about paganism to portray plants as sacred only through their *association* with a god/goddess. However, the sole focus on sacred plants associated with gods/goddesses does not effectively distinguish the relationships between plants and people in pagan worldviews. It also tells us little about the nature of the plants themselves and fails to properly distinguish the pagan and Christian attitudes toward plants. While Christians cut sacred groves as a demonstration of their superior faith, they too sanctified natural forms in the name of their own religion.[22] Like paganism, Christianity has its own sacred plants such as the grape vine, which are strongly associated with divinity in the Bible.

Constructing the idea of sacred plants only in terms of *association* with divinity depicts the plants themselves as *signs* or *symbols*, not as beings worthy of respect.[23] For a study of old pagan attitudes and behavior toward plants this approach is misleading. It fails to highlight that while biblical attitudes toward the natural world are hierarchical, in old pagan materials, nature is perceived as more of a heterarchy.[24] From the fragments of evidence that are available it is clear that the general worldview of pagan Europe involved the recognition of kinship between human beings and plants. Such expressions of kinship between humans and the natural world are a major source for contemporary pagan relationships with nature.

In an expansive study, Bruce Lincoln has demonstrated that connectedness between all parts of the natural world is one of the central themes of Indo-European creation narratives.[25] A common motif in Indo-European cosmology is the world coming into being through the death of a *first being*—usually a god, a man, or a livestock animal. Our previous discussion of Indian religions has highlighted such descriptions in the *Rig Veda*, but Lincoln claims that descriptions of a world emerging from the dismembered body of an original being pervade the extant cosmogonies of almost every Indo-European cultural group.[26] The most important aspect of such a cosmogony is that it does not rest on the level of the symbolic. It describes an actual and well-defined *consubstantiality* between everything of heaven and earth, which is itself the basis of relatedness and kinship.

Lincoln notes that the *Grimnismal* from the *Poetic Edda* expresses a similar theme to the *Rig Veda*. However, the relationships in this text are significantly more Earthly. Continuing the theme of the dismembered first being, the poem relates that the whole of creation is formed from the flesh of the dismembered Ymir:

> From Ymir's flesh   the earth was made
>   and from his sweat or, blood, the sea;
>   Mountains from his bones,   trees from his hair,
>   and heaven from his skull.[27]

Although the *Poetic Edda* was written down by Christians between the ninth and twelfth centuries, Bruce Lincoln's analysis suggests that the pervasive theme of consubstantiality is Indo-European in origin. Lincoln himself uses this passage as the basis of his theory for an Indo-European construction of homologies between specific body parts, such as flesh and earth. These homologies he terms *alloforms,* and he defines them simply as "alternative shapes of one another."[28] In the cases of flesh and earth, or of hair and plants, they can be seen as consisting of the same fundamental substance only superficially shaped into different forms. These forms are part of a continuous process of change "whereby one is continually transmuted into the other."[29] Plants and human beings are fundamentally of the same stuff. This shared substance between humans and plants is made explicit in another passage in the *Poetic Edda*. The *Seeress's Prophecy* relates that the first human beings were formed from the early green plants of the Earth. Odin, Hönir, and Lódur used the wood of the ash and the elm to create the first humans.[30]

A similar transformation of plants into human beings also occurs in the medieval Welsh poem *The Mabinogion*. Again, although this text is composed by Christians, the recognition of human-plant consubstantiality is very similar to much older pagan materials.[31] The tale *Math Son of Mathonwy* tells how Math and Gwydyon use magic to make a woman out of flowers. Using the flowers of oak and broom and meadowsweet, they conjure up a beautiful girl and name her Blodeuedd (from *blodeu* "flowers)." Later in the tale, the girl continues her transformation by being turned into an owl.

In the pagan Greek myths, the metamorphosis is often reversed, with human beings commonly transformed into plant forms. In a very similar way to the Indigenous materials, the transformations that take Daphne into the laurel, Minthe into mint, and Hyacinthus into a flower, show directly the shared substance of all living beings. These metamorphoses show how death links humans and other living beings; a feature shared with ancient Indian scriptures.[32] As *alloforms,* the current manifestations of living beings interchange at the time of death.[33] This process of transformation of one form into another not only demonstrates consubstantiality; it also uncovers a basic ancestral kinship between human beings, plants, animals, and the other inhabitants and constituents of the earth. All the plants, rocks, humans, and animals have the same genealogical ancestry. In the *Grimnismal*, plants and human beings can both trace their origins back to Ymir. Plants and humans are directly part of the same family.

This idea of common origin can be found in the Greek creation myth from Hesiod's *Theogony*. In this story of creation, rather than a dismembered being, it was the Earth that gave birth to everything, including all the gods, human beings, and plants.[34] The Earth also gave birth to the plant nymphs, or dryads, which in the *Homeric Hymns* appear to be inseparable from their plants:

> When the fate of death is near at hand, first those lovely trees whither where they stand, and the bark shrivels away about them, and the twigs fall down, and at last the life of the Nymph and of the tree leave the light of the sun together.[35]

This passage details an intimate relationship between plants and the dryads, with their lives intertwined and synonymous with the other. This relationship is expressed clearly in ancient Greek plant nomenclature. Amongst others, the cherry (Kraneia), the mulberry (Morea), black poplar (Aigeiros), elm (Ptelea), the grape vine (Ampelos), and the fig (Syke) were given the name of their nymph. All these dryads and all the plants were born from the Earth along with the gods/goddesses and human beings. They all share a common ancestry and are all broadly kin. With human beings capable of transformation into dryads, the existence of these nymphs strengthened the web of kinship between the human, the divine, and the plant realms.

Using this broad idea of kinship between humans and the rest of the natural world, several instances of more specific kinship links between humans and plants can also be found in material from pagan Europe. Predominantly, as in Indigenous animisms, there is evidence of specific kinship relationships constructed as an immediate familial connection. The prevalence of this Earth-based kinship between humans and plants differentiates the plant content in broadly animistic traditions from the material found in ancient Indian scriptures. As well as emphasizing connectedness, these relationships introduce notions of care, responsibility, solidarity, and deep appreciation. These familial relationships take the recognition of connectedness into the social sphere and toward action and responsibility. There is evidence for this in the classical myths of Europe.

In the sylvan myths of Rome, legend has it that Romulus and Remus's discovery beneath a fig tree was a result of tree parentage. It is thought that the milky sap of a fig tree suckled the twins after they were born. In the Roman world, the poet Virgil also describes direct kinship between humans and trees:

> In these woodlands the native Fauns and Nymphs once dwelt, and a race of men sprung from trunks of trees and hardy oak.[36]

This idea of tree parentage imbues these plants with the responsibility due to family. Similar ideas of tree parentage appear in ancient Greece with Hesychius who wrote of "the fruit of the ash the race of men."[37] In the vernacular Irish tradition, ideas of close relationships between humans and plants appear in names such as Mac Cuil, "Son of Hazel"; Mac Cuilinn, "Son of Holly"; and Mac Ibar, "Son of Yew."[38] A substantial connection with the Yew also appears in the Irish tale *Aislinge Oenguso* with the appearance of a character called Caer Ibormeith or "Yew Berry."[39] In Britain, comparable plant kinship can also be read in names such as in the old Welsh Guidgen, "Son of Wood"; Guerngen, "Son of Alder"; and Dergen, "Son of Oak."[40] According to Dowden, the old Irish name Ibor and the Gaulish Ivos (yew) "may possibly account for the tribal name *Eburones/Eburovices* and the town of York *Eboracum*."[41]

This emphasis on kinship and relationship can once again be characterized as an ethic of connection. The close, social, links between people and plants are very important as they both reinforce and complement the notion of consubstantiality found in the creation myths. Even though the evidence is fragmentary, it is clear that specific kinship ties helped form strong connections between particular plants and individual people within old pagan Europe. Pagan Europe's placement of plants and people in a relational heterarchy rather than a human-dominated hierarchy, demonstrates that a tradition of kinship and connection is not restricted to non-Western cultures. As Chapter 3 demonstrates, part of the reason for the loss of this ethic of connection is due to the marginalization of paganism in Europe with a Christian tradition that was founded upon the biblical, hierarchical view of nature. As part of a worldview that recognizes the connectivity between beings, and the responsibility due on the part of the human being, plants are not backgrounded or silenced in order to subjugate them as merely resources for human beings. By drawing on the recent animist scholarship, we can fully explore how plants fitted into the old pagan worldviews. The fragments of old pagan texts recognize these real, corporeal kin as subjective, aware, volitional, intelligent, relational beings.

Pagan Plant Persons

Lynn White Jr attempted a defense of paganism by portraying it as a religion that perceived nature imbued with spirits, whose "feelings" needed to be taken into account before nature could be used by human beings.[42] The discussion of animistic cultures has highlighted the error of connecting animisms with a naïve belief in nature spirits.[43] Such a characterization implies that animism is a failed epistemology.[44] Instead of spirit worship, interactions with the natural world are increasingly being theorized in terms of kinship links and relationships between *persons*.[45] The "old animism" of naïve belief in spirits is being replaced by a "new animism" that is based upon relating.[46] The premodern relationship with the natural world can, therefore, be predominantly characterized as *social*.[47]

Contemporary animism is mainly found amongst hunter gatherers, but animist views and practices are also present in agrarian and pastoral societies.[48] From the textual evidence already presented, in pre-Christian, agrarian Europe, there are animist tendencies. Plants and humans were regarded as kin, and there is also evidence that plants were also related to as persons—subjects to be related with as well as objects to be used. Thus, in a number pagan poems, myths, and songs (from across the multitude of pagan cultures), plants display their volition, agency, and subjectivity.[49] Plants are also widely depicted to be capable of communication with human beings. As in Indigenous animist cultures, it can be argued that the purpose of the stories and songs featuring persons is not necessarily to "explain phenomena in causal terms," but to learn how to recognize persons and how to find appropriate ways of relating with them.[50]

The mythology of the ancient Greeks is an appropriate starting point for a discussion of pagan plant persons. Although we are limited to the literary documentation of oral tales, in actuality, because the Greek tales were written down while they were still being communicated orally, they perhaps provide the best sources for exploring plant-human relationships within Europe. Once again, the Greek myths are notable for their use of motifs such as transformation, initial humanity, suffering, and blood to express plant personhood. As has already been discussed, the ancient Greek myths refer to the existence of plant nymphs, or dryads, regarded by Homer as being coextensive with their plants. Although not a traditional interpretation of the Greek myths, it is my contention that the existence of a dryad is a depiction of the plant as an other-than-human person.[51] As in Indigenous cultures, the dryads undergo transformation from a state of initial humanity to become the plants that they are synonymous with. As well as representing kinship, this initial humanity expresses subjectivity in the plant world.

In the corpus of Greek mythology, there are many instances of initial human forms being metamorphosed into plant life. Perhaps the most famous of these is the transformation of Daphne into the laurel tree, a tale which has inspired artists, writers, and poets for centuries. In this section, however, I wish to focus on two less famous examples of plant transformation. The first of these appears in the myth of the Heliades, documented by Virgil and his fellow epic poet Ovid. The sisters were the children of the sun god Helios, whose brother Phaëthon died recklessly. After the death of their brother, Diodorus Siculus relates:

> His sisters vied with each other in bewailing his death and by reason of their exceeding grief underwent a metamorphosis of their nature, becoming poplar trees. And these poplars, at the same season each year,

drip tears or sap, and these, when they harden, for what men call amber, which in brilliance excels all else of the same nature and is commonly used in connection with the mourning attending the death of the young.[52]

As well as shared substance, this transformation expressly recognizes personhood in the plant kingdom. This personhood is expressed by the continuing tears of pain and suffering emanating from the tree.

Initial humanity, subjectivity, and personhood are also motifs in the myth that deals with the origin of the myrrh tree and the birth of the famous Adonis.[53] In this myth, it is related that Myrrha was a beautiful princess who tricked her own father into an incestuous relationship and fled her country after being discovered. As a punishment, Myrrha's body is transformed into a tree, and yet she still weeps anguished tears, which take the form of the resin emerging from the myrrh tree.[54] The child she is carrying continues to grow within the tree trunk. When the child is born, the tree cries in pain. Not only does this myth relate the person-like nature of the tree, but it also contains another specific kinship link between people and plants. The mother of the human child Adonis is a tree.

As in the myth of Myrrha, a prevalent aspect of personhood is the depiction of trees and plants undergoing suffering in the same way as animals and human beings. In both Greek and Roman myths, kinship and suffering is expressed through the image of plants that bleed when cut.[55] An example of plants that suffer and bleed is found in the myth of Erysichton, a man who spurned the gods and refused to offer votive garlands and fragrant sacrifices to them. In several versions of this myth, Erysichton is said to have entered the sacred grove of Demeter, the Greek goddess of agriculture, with aggression toward the trees in mind. In the words of Callimachus:

> They rushed shameless into the grove of Demeter. Now there was a poplar, a great tree reaching to the sky, and thereby the Nymphai were wont to sport.[56]

Ignoring the rules of the sacred grove, Erysichton commands his men to cut the tree. Out of respect, his men refuse and force Erysichton to grab the axe himself. As Erysichton approached the tree:

> The sacred oak gave out a groan and shuddered
> and its leaves, its acorns and its branches paled.[57]

In this myth, the oak tree is clearly aware of its fate, and wants to avoid the loss of life that the axe of Erysichton will bring. Despite this communication from the tree, Erysichton hits the oak with his axe. As he does so, blood starts to pour from the bark. Erysichton however sticks to his task. Eventually this cascade of blood is coupled with even more obvious manifestations of pain and suffering. Ovid turns this act of person-person violence into a cautionary tale. Demeter sees that her trees have been in pain, and in revenge, she condemns Erysichton to a life of famine.

More important than the punishments are the portrayals of subjectivity, self-awareness, perception, and the ability of the tree to suffer at the hands of man. A similar portrayal of plant suffering through loss of blood is found in Virgil's *Aeneid*. Upon landing at Thrace, Aeneas requires some leaves to make a sacrifice to Venus, but when he attempts to uproot a myrtle tree, the tree lets out a cry and begins to bleed. Aeneas carries on regardless, only to find that the tree is his friend Polydorus transformed from human to plant.[58] Although Aeneas does not incur a punishment for the violation of trees, this portrayal of living, feeling plants fits well with Virgil's animistic world.[59] In Book 5 of the *Eclogues*, Virgil also refers to plants shedding tears, another common expression of awareness and suffering in trees.[60]

Here it must be clarified that these depictions of the ability to suffer are not due to primitive understanding of anatomy and physiology. Although suffering is not identified by Harvey as a necessary aspect of personhood, expressing its existence engenders the respect and consideration that is potentially lacking if plant are assumed to be insensitive to human action. From these Greek mythologies, it is demonstrable that some ancient Greeks had a sense that human action was capable of producing negative effects on plants. As in a human social context, knowledge of this subjectivity alone does not preclude the infliction of damage or death, but it is fundamental for the construction of respectful relationships.[61]

Such respectful relationships with plants are not limited to ancient Greek culture. There is also evidence for them in textual material across Europe. Comparisons between Old Norse/Anglo Saxon and the Greek myths reveal that as well as the Indo-European recognition of consubstantiality and kinship, these myths also contain depictions of plants as other-than-human persons.[62]

A passage from the *Poetic Edda*, describes aspects of personhood in relation to the world tree Yggdrasil. Like the perceptive trees of ancient Greece, Yggdrasil is depicted as a being that is capable of being harmed. The poem *Grimnismal* from the *Poetic Edda* lets it be known that this is not always appreciated by humankind:

> The ash of Yggdrasill suffers agony
> more than men know,
> a hart bites it from above, and it decays at the sides,
> and Nidhogg rends it beneath.[63]

As well as the capacity to be harmed and feel pain demonstrated by Yggdrasil, the remaining Anglo Saxon sources depict plants with other aspects of personhood. One example is the *Anglo Saxon Rune Poems*, which contain both practical and celebratory references to plants. The yew is lauded for its usefulness for firewood and building, and the poplar is thought to be splendid for its beautiful form. Likewise, the ash is praised as a tree precious to men, and in its resistance to the axe, this tree is recognized to possess a subtle agency and autonomy. Like any other living being, it wants to continue living:

> The ash is exceedingly high and precious to men.
> With its sturdy trunk it offers a stubborn resistance, though attacked
>   by many a man.[64]

While very subtle in the case of the ash, these person-like qualities are more strongly suggested in the poem for the oak:

> The oak fattens the flesh of pigs for the children of men.
> Often it traverses the gannet's bath,
> and the ocean proves whether the oak keeps faith
>   in honourable fashion.[65]

It is noticeable that in this poem the oak is subtly rendered both as a *subject* and an *object*, for it is the oak that *does* the fattening of the pigs and the traversing of the oceans and like other kinsfolk is expected to maintain its honor by keeping its people afloat at sea. As in contemporary animisms, it appears that the oak retains its personhood both in life and in death. In this way, as a living tree and as timber, the oak is a person that can be both "spoken about" and "spoken with."[66]

This recognition of relational autonomy is not restricted to the Rune Poems; similar notions also appear in the Anglo Saxon *Nine Herbs Charm*. Although the charm contains clear Christian references and was reportedly recorded by Christians either in the tenth or the eleventh century, the pagan recognition of plant personhood is clearly discernible.[67] The opening lines of the incantation are:

> Gemyne ðu, mucgwyrt, hwæt þu ameldodest,
> hwæt þu renadest æt Regenmelde.
> Una þu hattest, yldost wyrta.[68]

> Remember, Mugwort, what *you* revealed,
> What *you* established at the mighty proclamation.
> Una you are called, oldest of herbs.[69]

Within these few lines, the depiction of mugwort (*Artemisia vulgaris* L.) as a subject is clearly evident. In the *Nine Herbs Charm,* the mugwort is represented as capable of acting and remembering, both of which are qualities of active, intelligent beings. The opening of the charm also demonstrates dialogical communication between human beings and plants. The herb is purported to reveal something to the human, probably through ingestion as an aid to lucid dreaming, and the incantation is actually addressed toward this useful and powerful herb. With the repeated use of the word *þu* (you), the mugwort is treated not simply as a useful object, but also as an other-than-human person.[70] Perhaps this is reflected in a number of common English (person-like names) for mugwort—including Old Man, Naughty Man, and Old Uncle Henry. This recognition is not solely the preserve of mugwort, of the other nine herbs in the charm; chamomile and plantain are also addressed in this fashion:

> And, you, Waybread [Plantain], mother of herbs,
> open to the east, mighty within;
> carts rolled over you, women rode over you,
> over you brides cried out, bulls snorted over you.
> All you withstood then, and were crushed;
> So you withstand poison and contagion.[71]

In this passage addressed to plantain, the person-like characteristics expand beyond the use of the personal pronoun *þu* to a depiction of the plant as a flourishing being. The plantain is described as having an inner strength, which is directed toward resisting damage by humans and animals. In this charm, the plantain is *striving purposefully* to maintain its integrity when under physical threat, in the same way as other living persons do.

As well as Greek, Norse, and Anglo Saxon materials, depictions of plant personhood also occur in the mythologies of Western Finland. Indeed, some of the most complete and interesting European accounts of plant personhood occur in the Finnish epic poem the *Kalevala*. This is an epic presentation of a collection of traditional oral poetry, which recounts the creation of the remote Finnish province Karelia. One of the most interesting aspects of the *Kalevala* is that it is based upon traditional poetry collected in the nineteenth century from this remote area of Eastern Finland. Like the *Kalevala,* the original poems are animistic and mythical in nature.[72]

Of interest to this study is the fact that the *Kalevala* depicts plants as persons in very similar ways to both the Norse and Greek materials, which we have already discussed. One of the most striking aspects of the plants that populate the *Kalevala* is their ability to engage in dialogical interaction with the human beings that they come into contact with. The stories of these plants are important for bringing alive the qualities of personhood for the people who know them and hear them, thus allowing human beings to relate more dialogically with the world around them.[73]

In a number of passages from the *Kalevala*, plants are depicted as independent beings with their own perspective on life and their own "voice" for expressing this perspective. A good example of this relational nature occurs in a passage when the hero Lemminkäinen's mother searches for her son who is lost in the underworld.[74] In the depths of the forest and in need of help, she asks the trees about her son's whereabouts, and in response we discover:

> A tree talked, a fir tree sighed
> an oak skilfully answered,
> "I have worries of my own
> without worrying about your son. . . ."[75]

In these few lines, the oak displays its awareness and voice. Even though it is potential firewood for human beings, the oak's perspective on the world isn't muted. The story allows the tree to speak. In contrast to the biblical portrayal of plants as nonliving, passive resources, the tree is cast as a vital, relational, and intelligent subject. Like for Buber, the tree is a "Thou" rather than an "It."[76]

In addition to these communicative aspects of personhood, the *Kalevala* echoes the ancient Greek myths in relating the capacity of a birch tree to suffer. In the relevant passage, the epic's principal hero Väinämöinen enters a forest opening and hears a birch tree crying. Väinämöinen empathizes with the tree's pain and attempts to comfort it by reassuring it that it will never have to face the horrors of battle like a human being. The birch, however, speaks to Väinämöinen and relates its own worries. The poem expresses that the tree is a being capable of being harmed:

> Woe is me, I dread
> having my bark stripped
> my leafy twigs taken off![77]

Balancing Violence and Kinship

Old pagan sources also demonstrate that the recognition of kinship and personhood in plants does not preclude the use of plants for human needs. In the same ways as living peoples, ancient pagans needed to violate the autonomy of plants in order to provide themselves with food, clothing, fuel, and shelter. In pagan Europe, plant kindred would have been regularly killed for the benefit of human beings. The poems of the *Kalevala* show directly that relating to plants with respect does not prevent them from being damaged or killed for human needs. This is particularly the case in a harsh environment such as the eastern forest of Finland, which regularly requires the cutting of trees for human survival.[78]

In its opening passages, the *Kalevala* describes the destruction of the forests to make way for agriculture, and the hero Väinämöinen is forced to cut down a gigantic oak, which is blocking the sun and hindering the other forms of life on Earth. There is no attempt to deny that using plants involves killing persons. An example of predatory interaction with plants occurs when Väinämöinen requires wood to build a boat. To find the wood he needs, he sends the boy Pellervoinen off into to the forest. The first tree that he reaches with his axe shows that it is aware of the boy approaching for wood and intelligently asks:

> What, man, do you wish of me—
> what anyway do you want?[79]

Pellervoinen considers his position and tells the aspen that he is looking for wood to build a boat and requires the consent of the tree before chopping it. In reply, the aspen displays its self-awareness and craftiness, by saying:

> Full of leaks a boat from me
> And a craft likely to sink!
> I am hollow at the base.[80]

So the aspen list his faults and dissuades the boy from cutting him down. Pellervoinen moves on to a fir tree, which gives the human a similar reply:

> Not from me a craft will come
> One that bears six ribs!
> I am a gnarled fir.[81]

Finally the boy moves on to an old oak tree, and he asks the oak directly if he would be happy to be made into a boat. This time the oak displays self-awareness by recognizing that he has just enough wood to be made into a boat. The oak assents to the boy's request, and although the passage does not describe the oak suffering from the axe, the whole passage makes it clear that Pellervoinen is killing a tree that is alive and perceptive. Although Hutton may doubt the differences between pagan and Christian approaches to the natural world, such a narrative understanding of plant life contrasts sharply with the account of plants in the Bible.

Contemporary Paganism

These expressions of personhood and kinship in old pagan sources have served as inspiration for contemporary Pagans seeking to reestablish care-based relationships with other species. The revival of Paganism in Britain can be traced back to the 1700s, but contemporary manifestations of Paganism began to take shape in the latter half of the twentieth century. One of the three basic principles of the Pagan Federation includes the expression of a "love for and kinship with nature."[82] This sense of humanity situated in a heterarchical natural world closely mirrors the depiction of consubstantiality and relatedness found in the old pagan texts. Lacking in a continuous living tradition of cultural practices, contemporary Pagans have drawn on old texts as a basis for a contemporary Earth-based religion. But just as important as incorporating ideas from old texts, are visceral experiences of nature, direct experiences of nature's autonomy. As the Order of Bards, Ovates and Druids (OBOD) explains:

> Every part of nature is *sensed* as part of the great web of life, with no one creature or aspect of it having supremacy over any other. Unlike religions that are anthropocentric, believing humanity occupies a central role in the scheme of life, this conception is systemic and holistic, and sees humankind as just one part of the wider family of life.[83]

In this respect, an appreciation of connectivity is not just based on textual sources, but crucially on close personal experience and observation. In turn this appreciation is based upon a will (or indeed a need) to celebrate the natural world, to abandon domination, and to refind connections with others. In conversations with Pagans about plants, many will talk about the "spirit" of a herb or tree. Damh, from OBOD gave me her views on human-plant relationships in Druidism:

> We don't tend to view them as inanimate life forms, but rather as life forms with their own consciousness and spirit.[84]

If we are to avoid epistemological and ontological tangles, I agree with Harvey that "words like 'soul' or 'spirit' are not always useful, but are attempts to say what it is about something which makes it alive."[85] In my view, it also an attempt to say that plants are not passive, unperceptive beings—that plants are autonomous and share the subjective experience with human beings. This shared experience allows dialogical interaction. As in contemporary animistic cultures, many contemporary Pagans maintain that such dialogue between people and plants involves the transfer of information. As Damh explains:

> By holding such a view of life [of kinship] it is then possible to open up a deeper relationship with a plant, through communication with its spirit.[86]

Although not all talking to trees yields factual communication, maintaining dialogical relationships with plants always acknowledges their presence and voice.[87] An example of such dialogue in action is the practice of the "beating of the bounds" which takes place at camps organized by the OBOD. The ceremony is intended as a greeting toward all the other-than-human beings who live in the vicinity (or in the middle) of the camps. It is an acknowledgement of the personhood of the other-than-humans through a greeting in human language. Such "talking to plants" is a common trait in people who work closely with plants, and whether done consciously or not, it is a way of acknowledging their presence.[88] Fundamentally, this is a method of building relationships with plants that are not based exclusively on their instrumental value to human beings. This Pagan openness to trees in particular can lead to Pagans giving human voice to the plant persons who they are in dialogue with:

> I have thrust my feet into the dark, rich earth and fed,
> I have drunk from the crystal waters of the depths,
> I have clothed myself anew, in shades of green, to greet fair Spring,
> I have been washed clean in Summer's rain,
> I have shed my raiment, now gold and scarlet, at Autumn's command,
> I have stood proud and naked before Winter's sun.
> I have housed a million lives within my boundaries,
> I have watched lovers embrace at my feet,
> I have comforted the lonely child in my arms,
> I have been cut down and my body used to bring warmth,
> I have grown once more in the cycle of life,
> I have been one with the Forest, yet a single being.
> I am the Tree.[89]

The use of stories or poems to display the voice of plants is aimed at finding dialogue between Pagans and their environment. In this way, bringing the animated presence of plants to the foreground helps to avoid the "monological idea that I am responsible for reconnecting myself to the world."[90] Pagans often relate to plants as persons by regarding specific local trees as friends to be greeted or by regarding local species as ancestral. Echoing the specific kinship relationships found in old European Paganism, Pagans in Britain have informed me of their particular kinship relationships to trees. In the conversations that I have had with Pagans, a man from near the Chiltern Hills expressed his reverence for the local flora:

> Sitting in the woods is sitting with your ancestors. Trees are necessary for life—they provide shelter, fire and food—but they are also family members. In my local woods, each tree is unique; each has its own characteristics and personalities. To me the beech is a mother and the oak a father; and knowledge of that connects me with my human ancestors and their beliefs and practices.[91]

Again, this understanding of an entwined human-plant kinship is achieved through direct experience of, and an open, detailed appreciation of the plants in question. It is important to recognize that the reconstruction of these local, specific kinship relationships with plants is the result of an intense need to step beyond the largely utilitarian approach to plants that characterizes mainstream Western practice. Instead, contemporary Pagans actively seek relationships of care and responsibility for their kin.

Perhaps the most visible displays of such care occurred during the British antiroad demonstrations of the 1990s. These saw a large number of eco-Pagans working to protect a number of ancient broadleaf woodlands threatened by a widespread road-building program.[92] The acts of defiance that characterized the construction of eco-Pagan camps were aimed at protecting a large number of local woodlands. Within these woodlands were plants and animals that were the subjects of personal relationships with local Pagans who were forced to act.[93] Although popular media characterizes such care as "tree hugging," Pagans recognize that relationships with other-than-human persons are not always nice. Like other animists, indeed like all other humans, Pagans must have violent relationships with plants. As one Wiccan practitioner writes, "I see plants as people, too, and know that something must die so I can survive."[94] Plants must be subjected to violence for Pagans as much as for anyone else, but like other animists, many Pagans recognize this violence and seek ways in which to mitigate it. Pagan lessening of violence toward plants can be expressed in small lifestyle changes, such as avoiding food and paper wastage or leaving patches of garden for the benefit of local, nonhorticultural plants. Small acts such as avoiding the cutting of live trees are also encouraged by Pagan writers.[95] These are similar expressions of respect to those shown in Indigenous animistic practices.

Like other animists, where violence is deemed necessary, many Pagans will express respect for the autonomy and personhood of plants by *asking* plants for permission to take leaves, stems, and roots. While it is difficult to determine whether any plant assents to be killed, this simple act is recognition of the fact that plants are not solely for human beings to use. Gratitude, respect, and appreciation of the role of plants in sustaining human lives are expressed by making offerings to plants, either individually or as part of a Pagan group. Barry Paterson recommends offering human blood to trees that must be cut as an expression of solidarity and kinship.[96] Other Pagans make physical offerings of water, alcohol, tobacco, and nonphysical offering of prayers, love, and compassion. Such ritual acts are forcefully encouraged by Ronald Grimes as a way of deeply identifying with all the Earth's creatures to the point that human beings lose their arrogant sense of superiority.[97] While rationalist sceptics may question the value of such behavior (especially for the plants themselves), the accumulation of such acts serve to resituate the human being in a constructive relationship with the plants which surround them. And, as Graham Harvey points out, small acts pave the way for larger ones.[98]

Yet, as the road protest movement showed, many Pagan groups are also directly involved in larger activities that seek to restore responsibilities of care to other-than-human persons. Many Druid groups—such as the Pagan Federation, The Druid Network, and Dragon Environmental Network—partake in conservation projects and the restoration of local woodlands. The website of The Druid Network states the importance of conservation activity to Pagans: "a significant part of The Druid Network's work is raising funds for the planting of trees and conservation of ancient forests."[99] Although some environmental writers have questioned the value of such intervention, for Pagans, such action attempts both to rectify and lessen the human violence toward plants, animals, rocks, and other beings.[100] Even though plants may be "planted" by humans, this does not vitiate their autonomy. Instead, it establishes a significant care-based kinship partnership between human and plant. In its own way, this helps overcome the backgrounding of plants both planted and free-living, because ultimately for many Pagans:

> The plant is a living, intelligent, conscious being. We don't use plants as if they were just substances or tools. Instead, we form a type of relationship, just as you might do with friends.[101]

Stemming from personal experience, such attitudes demonstrate that there is not a single Western attitude to plants. Not all Western approaches to plant life are aimed at backgrounding them for domination. While such claims are true of contemporary Paganism, remarkably, they are also true of the emerging perception of plants in the Western plant sciences. The following chapter looks at how the growing appreciation of plants as intelligent, active beings stands in sharp contrast to the commonplace idea of plants as vacant spaces. Indeed, the view of plant scientists is changing so rapidly that botanical texts may soon serve Earth-based religions for inspiration as much as the pre-Christian scriptures.

7

BRIDGING THE GULF

Moving Sensing, Intelligent, Plants

> Our view of plants is changing dramatically, tending away from seeing them as passive entities, subject to environmental forces and organisms that are designed solely for accumulation of photosynthetic products.[1]
>
> —František Baluška

Claims of a constructed human-nature separation have to acknowledge that within scientific circles, since the publication of Darwin's *Origin of Species*, humans and plants have been recognized as sharing a common (if distant) ancestor. This situation has been addressed by Plumwood, who writes that the "insights of continuity and kinship with other life forms . . . remain only superficially absorbed in the dominant culture, even by scientists."[2] In spite of scientific knowledge of relatedness, the natural world and the plant kingdom remain backgrounded. Plumwood asserts that this domination of the natural world is perpetuated by "continuing to think of humans as a special superior species" and consequently other species as inferior.[3]

In such a philosophy of exclusion, the identity of the superior group must rest on the constructed inferiority of others. This rendering of plants as radically different and inferior helps maintain the dominant human sense of collective superiority. Despite awareness of kinship, this denigration of plant ontology has a large role to play in the human elevation above, and separation from, the wider natural world.

Significantly, this positioning of plants as radically different—as passive, insentient, inferior beings—is contradicted by an overwhelming body of evidence that has been accruing in the botanical sciences. From close observations of plant behavior, botanical science has implicitly rejected the zoocentric inferiorities of plants such as lack of movement, lack of sensation, and lack of mentality. Indeed over a period of approximately two hundred years, plant science has built up a bulk of evidence that shows plants in a very different light. In many ways, this chapter can be considered as a synthesis and interpretation of botanical evidence, which aims to convince the Western mind of the sentience and intelligence of the plant kingdom.

The foundations of this evidence are based upon historical advances in plant anatomy during the seventeenth and eighteenth centuries—such as the discovery of cells in plants by Grew and Malpighi, Camerarius's recognition of plant sex, and Hales's breakthrough on plant respiration.[4] While these discoveries helped to bridge the gulf between plants and animals, this chapter focuses on evidence that refutes the notion of plants as passive, nonmoving, insensitive, nonminded beings.

Darwin not only put forward the idea of relatedness between humans and the natural world, but his work was the first to fully demonstrate and articulate the idea that plants are capable of movement and sensation—providing the basis for the discipline of plant signalling. Rather than studies of evolutionary biology, plant signalling is crucial because scholars within this discipline have begun to recognize many points of continuity in the natures and capabilities of plants and human beings. Here my aim is not to dispense a complete botanical history, but to highlight some key advances in our discussions of plant ontology and epistemology.

Sensation and Movement

The recognition of sensation and movement in the plant kingdom can be traced back to 1824 when Henri Dutrochet proposed the idea that the growth responses of plants to light was a *behavioral response*, not simply mechanical movement.[5] This claim was repeated in 1868 when Albert Frank put forward the idea that the responses of plants to gravity and light were *induced*. Rather than the mechanical actions of an automaton, Frank's hypothesis was that the growth movements of plants were active, coordinated responses to sensed stimuli. In 1878, this active sensation and movement was proven for the first time in experiments by Von Wiesner, which showed that plants continued to move toward sources of light, even after these sources had been removed.[6] These experiments demonstrated that a biological response process had been initiated; a response which carried on even after the stimulus had been turned off.

This research laid the groundwork for Charles Darwin's investigations into tropic movements in plants. It was Darwin's work that firmly established in the botanical sciences the existence of initiated, nonmechanical movement in plants.

In a long series of experiments, Darwin confronted plants with various stimuli and studied their subsequent growth and movement. In his *Power of Movement in Plants* (1880), Darwin carefully describes plant growth movements and draws attention to the remarkable similarities between the movements of plants and animals:

> But the most striking resemblance is the localisation of their sensitiveness and the transmission of an influence from the excited part to another which consequently moves.[7]

Darwin's observation of a transmissible substance is of great importance. The proposition that such an influence may exist, acknowledges the interconnection between the parts of plants and their synergistic integration as a whole organism. It is one of the first recognitions of internal signalling and communication processes in plants. At the same time, similar research was being conducted in animals, and by noting them together, Darwin built upon the work of earlier physiologists to further close the gap between the perceived abilities of animals and plants.

Darwin was particularly impressed with the action of the radicle tip (embryonic root). In the *Power of Movement in Plants*, he notes the ability of the radicle to sense (and move away from) objects that might elicit tissue damage. Darwin was also amazed at the ability of the radicle to grow actively toward sources of water and gravity. From such painstaking observations, Darwin recognized the capacity of plants to sense and choose, and attributed brain-like characteristics to the root tip:

> In almost every case we can clearly perceive the final purpose or advantage of the several movements. Two, or perhaps more, of the exciting causes often act simultaneously on the tip, and one conquers the other, no doubt in accordance with its importance for the life of the plant. The course pursued by the radicle in penetrating the ground must be determined by the tip; hence it has acquired such diverse kinds of sensitiveness. It is hardly an exaggeration to say that the tip of the radicle thus endowed, and having the power of directing the movements of the adjoining parts, acts like the brain of one of the lower animals; the brain being seated within the anterior end of the body, receiving impressions from the sense-organs, and directing the several movements.[8]

From his observations of the radicle, Darwin was the first person in the history of modern botany to recognize intelligent, purposeful movement in the plant kingdom. In Darwin's own lifetime, the investigation of tropic responses began to suggest that many of these purposeful movements were aimed at satisfying the nutritional needs of plants.[9] For centuries, botanists had also been aware of the so called nastic movements in sensitive plants such as *Mimosa pudica*.[10] Further work on these movements, again conducted by Darwin, led to the discovery of nervous impulses in plants. This is fundamental to contemporary evidence, which contradicts the notion that plants are passive.[11]

Darwin systematically studied the existence of nongrowth movements in *M. pudica, Drosera rotundifolia* L. and *Dionaea muscipula* Sol. ex Ellis (Venus Flytrap). This was aided by the first recording of electrical movement [or an action potential (AP)] in the leaves of *D. muscipula* by Burdon-Sanderson in 1873, who worked closely with Darwin.[12] In *The Power of Movement in Plants* and *Insectivorous Plants*, Darwin provided experimental evidence of the movement of leaves in these species. In *Insectivorous Plants,* Darwin investigated the movements of leaves that allow the capturing of small insects. Working with the sundew, *D. rotundifolia*, he elicited movement by various mechanical and chemical means and described the passages of "motor impulses" through the cells of the leaves and tentacles that enabled the plant to exercise movement.[13]

Darwin demonstrated that plants were able to perceive minute quantities of chemicals that he had administered to them. He considered this a remarkable occurrence. It led him to compare this perception and movement in plant leaves to the sensory capacities of animals:

> These nerves then transmit some influence to the brain of the dog, which leads to action on its part. With *Drosera*, the really marvellous fact is, that a plant without any specialised nervous system should be affected by such minute particles; but we have no grounds for assuming that other tissues could not be rendered as exquisitely susceptible to impressions from without if this were beneficial to the organism, as is the nervous system of the higher animals.[14]

Darwin recognized the possibility that plants could receive impressions of the environment. Although he was not aware how, it was clear that plants were able to communicate with the environment, and the sensory parts had the means to communicate this information on the state of the environment to other parts of the plant. Again, for this type of communication, Darwin posits the existence of an influencing substance. It is interesting to note that this position is in direct contrast to Aristotle who denied plants the ability to communicate with the environment because he could not fathom the means by which they were able to receive sensory impressions.

However, despite Darwin's views, the faculties of sensation and awareness in plants were not proven beyond doubt. Even though there was experimental evidence for sensation and movement, the sensitive plants *Mimosa, Drosera* and

Dionaea were regarded by the majority of scientists as unusual cases. Other plant species, which did not demonstrate nastic movements, were still regarded as passive.[15] Despite his ideas on sensory impressions being received internally and on the intelligence of the radicle, Darwin himself also played down his findings:

> Yet plants do not of course possess nerves or a central nervous system; and we may infer that with animals such structures serve only for the complete transmission of impressions, and for more complete intercommunication of the several parts.[16]

Unfortunately in Darwin's caveat, the zoocentric influence of Plato and Aristotle persists. Again the anatomy of the plants in question is judged in relation to the anatomy of animals. As plants do not share the complex tissue structure of animals, Darwin assumed their capacities to be in some way incomplete and lacking. Although he had demonstrated that plants sense and move in intelligent ways, this final caveat served to reduce the existence of plant perception to a simple vegetable level. Despite this, his work firmly established the notion that plants are capable of initiating movement and cemented previous experimental evidence that plants could perceive their environments through touch and through direct perception of light, water, and temperature.

Movement and Signalling

The ideas of Darwin on the tropic and nastic movements of plants and the presence of a signalling process were initially dismissed by the majority of plant physiologists. As visible movement and sensation had only been demonstrated in exotic sensitive plants like *M. pudica,* the general position that plants were insensitive was retained.[17] Sensitive species were considered somewhat anomalous to the common vegetable, and so the findings from studies on sensitive plants did not serve to change the perception of common plants. It is a measure of the depth of the acceptance of the mechanistic position and of plants as inferior beings—that as this evidence uncovered the existence of sense and self motion, these ideas remained entrenched.

Darwin was a pioneer in the field of the communication *within* plants by molecular and electrical signalling processes. Focussing on the transmission of electrical impulses in plants, after the work of Darwin was published, Haberlandt continued researching the phenomenon. He discovered that vascular tissues, in particular the phloem cells, facilitated the transmission of APs and concluded that if plants were to be considered as having nervous tissue, then the long phloem cells were the likely location. Some researchers thereafter referred to the phloem cells as "plant nerves," but the majority of papers and textbooks affirmed the belief the plants had no "nervous system."[18] This ignorance of Haberlandt's work on electrical conductivity in phloem tissue remains to the present day, especially in popular science. Even recent, sympathetic publications have taken on this misunderstanding and claimed that "trees have no brains or nerves and instead run their entire lives with the aid of a remarkably short shortlist of chemical agents."[19] Such interpretations ignore the fact that the nerve cells of animals and phloem cells of plants "share the analog function of conducting electrical signals."[20]

In the 1960s, another threshold was crossed in the understanding of the sensitivity of plants. Although it was a hardly noticed event, more run-of-the-mill plants than the exotic sensitive plants were discovered to conduct electrical signals.[21] If pumpkins produced APs, then the perception that sensitive plants were an exception to the rule of passivity was shown to be erroneous. This finding also encouraged biologists to believe that widespread electrical messaging must have a strongly adaptive function. As a consequence, electrical signalling has been found to have a role in vital processes such as photosynthesis, respiration, phloem transport, and systemic defense.[22] In addition to APs that occur in animals, higher plants have been recently found to use a unique long-distance electrical signalling method. This method is called the "slow wave potential."

As well as providing the foundations for work into electrical signalling, the early work of Darwin paved the way for research into plant molecular signalling. Darwin had demonstrated that the site of light perception in a shoot was at the tip, but that the location of the curvature was separable. He proposed that a transmissible substance from the tip was communicated to the region of curvature.[23] In 1931, investigations of this conjecture yielded the the hormone auxin, which is vital for tropic movements in plants, and this discovery stimulated widespread research on tropic growth and plant signalling.

Contemporary research into plant growth and communication has built upon the platform provided by Darwin. Studies in communication and signalling have shed light on the movements of the sensitive plant *M. pudica*. It is now known that the activation of a receptor on the leaves of *M. pudica* triggers an increase in intracellular calcium, which may act as the signalling molecule along with electricity. A signalling and communication process is indicated by the fact that not only do the touched leaves close, but leaves away from the source of the stimulus may close as well. Although the role of APs in plants is still poorly understood, the closure of *Mimosa* leaflets is achieved through a process of *sensing, communication,* and *action.* This response to stimuli is now thought to be ubiquitous in the plant kingdom. In a review of plant responses to stimulation, biologist Janet Braam makes it clear:

> From the violence of tree strangling and insect trapping to the elegance of roots navigating through barriers in the soil, responses to mechanical

perturbation are integral features of plant behavior . . . probably all plants sense and respond to mechanical forces.[24]

Touch remains the most well-known sense in plants and that to which biologists currently assign most importance in the sensory repertoire of plants. Roots in particular are extremely sensitive to touch, which "enables them to explore, with an animal-like curiosity, their environment in a continual search for water and solutes."[25]

However, we would be limiting ourselves if an explanation of plant sensory abilities ended with touch. As the phototropic response to light clearly shows, plants are able to directly perceive light. Plants use the perception of light to direct movement. They are also able to use light to sense the proximity of neighboring individuals, which may be future competitors. Measuring an increase in Far-Red light (reflected by green tissues), plants use this information to perceive their neighbors and to predict whether they will render them subject to shading. If shaded, complex, morphological shade avoidance responses ensue.[26] As we rrecognize the perception of light by the human eye as sight, in their own way, plants also "see" their neighbors. This vision allows plants to make decisions about the future, related to branching and flowering behavior.[27]

PLASTIC PLANT INTELLIGENCE

Founded upon Darwin's work and the development of signalling, a significant number of studies on a wide range of plant species have begun to move beyond demonstrating that plants are simply capable of sensation. Contemporary research in the plant sciences is demonstrating that plants possess many attributes of an active intelligence. Fundamental to this is the concept of *plant intelligence*, proposed in 2002 by Anthony Trewavas.[28] In the context of this study, the concept of plant intelligence is significant because it is an intentional attempt to discredit the notion that the Earth's most abundant form of life is passive and mechanical.[29]

Trewavas's description is founded upon Stenhouse's definition of intelligence as the possession of "adaptively variable behaviour within the lifetime of the individual."[30] While animals may behave by moving around from place to place, Trewavas points out that plants behave by movements in a particular place; movements which usually are the result of growth. The growth and development of plant organs is "adaptively variable." it changes according to environmental conditions in order to maximize fitness.[31] Therefore by definition, this ability to alter the phenotype is intelligent.

Although Trewavas has most recently pointed out the significance of this adaptive behavior in plants, it is not a recently observed phenomenon. For centuries it has been known to botanists as *phenotypic plasticity*—the ability that plants have to change their outward form in response to changes in the environment. An early description of the variation produced by phenotypic plasticity in plant species is found in Linnaeus's *Critica Botanica*. Linnaeus notes the appearance of several aquatic species, including a species of *Ranunculus* which:

> Put forth under water only multifid leaves with capillary segments, but above the surface of the water later produce broad and relatively entire leaves. Further, if these are planted in a shady garden, they lose almost all the capillary leaves, and are furnished only with the upper ones.[32]

Plants accomplish this plastic development in a way that by necessity involves assessment of the prevailing environmental conditions and the selection of appropriate responses.[33] Plasticity therefore can be regarded as the manifestation of a plant's awareness of the environment. The resources are assessed, and the most beneficial growth and development response is induced in the *whole* organism.[34] The existence of plasticity is actually vital for the survival of plant life, and Trewavas considers it "a visible witness to the complex computational capability plants can bring to bear to finely scrutinise the local environment and act upon it."[35]

The existence of plasticity demonstrates that "the behaviour of plants is not pre-programmed"—an assertion that contradicts the concept of plants as *automatons*.[36] Plants do not always operate predictably like a piece of clockwork. Instead, perception, awareness, and active assessment are crucial elements in the behavioral repertoire of plants. As is commonly recognized in animals, this intelligent, plastic plant behavior is directed toward an increased well-being through the optimal acquisition of resources and the maximization of reproduction. The ability to adapt to new and changing conditions typifies the intelligence of phenotypic plasticity. There are many examples of this, but here I will concentrate on the action of roots, because plant roots are perhaps the most plastic of organs and are under tight control by the organism as a whole.[37]

Rich soil patches are exploited by increased plastic root branching and root growth. In the presence of few nutrients, root growth has been found to accelerate in order to facilitate the detection of new, more nutritious patches of soil in other locations.[38] There is clear and active perception of the resources available, which for Trewavas involves the construction of a "three dimensional perspective" of the local space.[39] Here plants display their behavioral intelligence with an ongoing assessment of the costs and benefits involved in exploiting the resources that exist in the soil.

Plants clearly and intentionally avoid areas with poor nutrient levels. The active, below ground assessment and discrimination of soil resources is integral for plant nourishment and survival. Studies estimate that it can increase the absorption of essential nutrients by between 28 percent and 70 percent.[40] Root plasticity allows plants to make choices about the soil patches they feed in—to the extent that plants have been referred to in ecological studies as "foragers."[41] From close observation of plant behavior, therefore, it is apparent that plants use assessment mechanisms in a similar way to animals and explore the soil to optimize the gathering of food resources.[42] This perception and assessment also allows plants to avoid competition. The roots of certain desert shrubs have been found to use root plasticity to deliberately avoid contact and competition with roots of other species.[43]

The concept of plant intelligence has not been without controversy in the plant science community. It has been challenged by Richard Firn through the argument that as intelligence is a property of individuals, plants cannot be intelligent as they are not individuals in the same way as animals.[44] While Firn claims that the organs of plants operate individually, Trewavas cites substantial evidence to the contrary, demonstrating a remarkable amount of communication and cooperation between plant organs.[45] On the evidence of such communication and synergistic action, plants are clearly individuals.

The theory of plant intelligence has also been attacked by Firn on the basis that plant behavior is the result of machine-like reflex reactions.[46] Doubts about the theory of plant intelligence have also emerged in the plant science community due to its associated notions of reasoning, learning, and problem solving. Struik et al. claim that such ideas "inevitably invoke the notions of consciousness and free will, elements that are totally unnecessary if adaptive responses are considered passive as in a Darwinian world."[47] Not only does the idea of plant intelligence have little to do with notions of consciousness and free will (see next section), in these reductive dismissals of plant intelligence, there is the now familiar a priori assumption that plant activity is passive.

Do Plants Learn to Reason?

In order to deal with an enormous range of environmental conditions that have the potential to change rapidly over short spaces of time, Trewavas highlights that learning is necessary for plant life.[48] Like other forms of learning, this involves continual assessment and the ability to make behavioral corrections in order to reach a required goal. One of the most interesting of these experimental examples involves testing the growth abilities of plant tendrils. Growth experiments on plant tendrils have demonstrated their ability to assess the position of a support and actively move toward it. The immediate goal is to reach the support, but if its position is moved, then the tendrils are able to sense this change and adjust the direction of their growth movement in order to relocate it.[49] Rather than being involved in automated, repetitive, purely stimulus-driven behavior, plants make *real time* assessments of stimuli and respond according to both their current state and previous experience. At each point in a behavior event, Trewavas notes that the plant is acting upon information from previous responses; a form of trial and error learning. There is also an integration of external information with knowledge of the internal state. Assessment of this assimilated information guides appropriate action. This is a process of *reasoning* which is directed at the optimization of factors that will maximize fitness.[50]

In addition to the experiments on tendril growth, the responses of plants to water stress have also been put forward as examples of learning and reasoning. For a plant to respond to drought, for example by the abscission of leaves, it must be able to assess the present level of water against the optimum supply level. Trewavas summarizes:

> The plant learns by trial and error when sufficient changes have taken place so that further stress and injury are minimised and some seed production can be achieved. The responses to water stress are modified by interaction and integration with other environmental variables e.g. mineral nutrition, age, temp, history etc and are therefore not reflexive responses. Clearly *decisions* are made by the *whole plant*.[51]

In adjusting their morphology in response to often rapid environmental changes, plants are capable of basic decision making, problem solving, and reasoning. Remarkably, in addition to these intelligent faculties, there is also some direct experimental evidence for the existence of intention and choice in plants.

A study of the feeding displayed by the nonphotosynthetic parasite *Cuscuta europaea* L. demonstrates that this plant makes choices when selecting a host. These are based upon the level of sustenance that *Cuscuta* anticipates that the host will provide. If the host is deemed to have insufficient capacity to provide essential nutrients (i.e. if the host is revealed to be lacking in nitrogen), after initially coiling its tendrils around the plant, the dodder will choose not to continue with feeding. Instead, it will uncoil and keep searching for another host, a host more suited to its dietary needs.[52] This is a case of a plant employing an optimal foraging strategy to ensure that it does not waste resources. A related study has demonstrated that the host perception in a close relative, *Cuscuta pentagona* Engelm. is mediated by volatile emissions from the host plant. The parasite uses the presence of volatile chemicals in the air to sense the position of the host before growing toward it. In another example of choice, several species were found to activate a feeding response, but when *C. pentagona* was given a choice of hosts it was shown to actively prefer the tomato plant.[53] These examples have emerged from work on parasitic plants due to the much simplified scenario of working with easily identifiable resource hosts. However, rather than relying on hosts, the majority of photosynthetic plants must forage for resources in soil.

Although explicit experiments have yet to be conducted on active rhizospheric choice, plant roots prefer to be located in resource rich patches in the same way as parasites prefer resource rich hosts.

Plant Brains

As befits their modular structure and the ability to grow from each of their modules, unlike animals, plants have no use for a centralized brain and/or nervous system. Instead of centralized brain tissue, a newly emerging field of plant science, dubbed "plant neurobiology," is suggesting that plants may actually have thousands of brain-like entities that are involved in the emergence of intelligent behavior. These entities are a type of tissue known as *meristems*. Current theories suggest that the meristematic tissue, located at the tips of roots and shoots, combined with the vascular strands capable of complex molecular and electrical signalling, may well comprise the plant equivalent of the nervous/neuronal system.[54] In a groundbreaking text *Communication in Plants*, Baluška et al. echo the pioneering work of Darwin:

> Each root apex is proposed to harbour brain-like units of the nervous system of plants. The number of root apices in the plant body is high, and all "brain units" are interconnected via vascular strands (plant neurons) with their polarly-transported auxin (plant neurotransmitter), to form a serial (parallel) neuronal system of plants.[55]

Rather than following Darwin's judgement that this plant nervous system is inferior to that found in animals, plant neurobiology researchers regard this decentralized assessment and response system to be the most effective for maximizing plant fitness.[56] Such a system is thought to enable decentralized behavior (i.e., growth), which allows plants to thrive in complex and everchanging rhizospheric environments.

It has been proposed that in the plant the meristematic "brains" may exert influence on the rest of the plant tissue by the transmission of signalling molecules such as the hormone auxin. Auxins are manufactured at the root and shoot apices, and it is thought that their movement is one method for allowing the transfer of information throughout the individual. It has been proposed that the end poles (cross walls of cells) are analogous to the synapse in animals.[57] At so called "plant synapses," vesicular transport of auxin moves this signalling molecule from cell to cell. Although the exact processes have yet to be uncovered, it has been proposed that this extracellular transport of auxin "exerts rapid electrical responses" across the plant synapse and "initiates the electrical responses of plant cells."[58] Whatever the pathway within the plant, communication can occur over long-distances, with information on the environmental and developmental state of the roots being transferred to the shoots—as in the case of stomatal closure during water stress. As well as auxin and electrical signals, plants produce and use a variety of neurotransmitter molecules to communicate from cell to cell. Dopamine, acetylcholine, glutamate, histamine, and glycine are all touted as potential signalling chemicals between cells.[59] Other complex communication molecules include protein kinases, minerals, lipids, sugars, gases, and nucleic acids. Trewavas has drawn attention to this complexity and notes that "from the current rate of progess, it looks as though communication is likely to be as complex as that within a [animal] brain."[60]

In response to some of the assertions of plant neurobiologists, Alpi et al. have suggested that the existence of *plasmodesmata* (microscopic channels, which traverse plant cell walls and enable transport and communication between cells) contradicts the idea of plant synapses and of auxin as a neurotransmitter, as their existence facilitates extensive electrical coupling, precluding the need for any cell-cell transmission of a neurotransmitter-like compound.[61] However, this criticism has been refuted by Brenner et al., who assert that although the exact pathways are still to be discovered, auxin *is* known to be transported from cell-cell and active, communicative plant behavior does take place.[62] Along with the exact mechanisms of electrical cell-cell coupling, they assert that investigating these transfers represents an exciting field of study for understanding plant signalling and behavior.

With thousands of meristems, a plant has potentially thousands of "brain units." It is proposed by advocates of plant neurobiology that plants integrate sensory information and make decisions based upon communication between a multitude of plant tissues such as the root meristems, interior meristems, and the vascular tissues. Barlow has pointed toward the involvement of the vascular tissue (xylem and phloem) in conveying APs from zones of special sensitivity to other regions of the plant—an "informational channel" involved in organismal organization.[63] Trewavas has proposed that the meristematic tissue, which runs throughout the plant, could be an integrative assessment and computational tissue, acting with sensory input from local meristems.[64] With active debate on this topic, it is still to be uncovered whether this internal communication systems are centralized, decentralized, or somewhere in between.[65]

The structural complexity of these communication networks within plants is of great interest for an understanding of the intelligent behavior that plants display. The eminent animal physiologist Denis Noble has recently argued that network-style interactions (like those found in plants), actually organize and direct the activity of *all* living beings. In *The Music of Life,* he disputes the view that a unitary, external mind or self controls and directs the activity of living organisms.[66] Against this Cartesian notion, Noble argues that it is decentralized communicative networks that heterarchically self-organize and direct living activity.

In Noble's view of *systems biology*, "there is no single controller." no single Cartesian mind substance, which is the director of living systems.[67] Instead, from a systems viewpoint, mental properties such as intelligence, reasoning, and choice are thought to emerge from the interactions of physiological networks of signalling and communication. As Evan Thompson puts it, the "emergent process is one that results from collective self-organisation."[68] These principles of heterarchical organization and the emergence of higher level properties are fundamental principles of systems biology, which are elegantly summed up by Fritjof Capra:

> According to the systems view, the essential properties of an organism, or living system, are properties of the whole, which none of the parts have. They arise from the interactions and relationships between the parts. These properties are destroyed when the system is dissected, either physically or theoretically, into isolated elements.[69]

Although the exact pathways are still being investigated, we can state that from a systems perspective, the interconnecting, heterarchical network of plant tissues (including meristems) enables intelligent plant behavior, rather than the Cartesian consciousness or free will alluded to by Struik et al.[70]

Plant Self

Internally, plants rapidly exchange detailed environmental information between different organs. Although somewhat decentralized, a consequence of this integrated communication system is that plants also recognize themselves as integrated beings. In other taxa, this phenomenon is known as *self recognition*, and its existence in plant species is being remarkably demonstrated by experimental evidence.[71]

Through studies of breeding systems, it is known that plants are able to recognize self/nonself with some degree of precision. During pollination, the majority of plant species employ "self-incompatibility" mechanisms. Self-fertilization (and, hence, potential loss of fitness) is prevented by discriminating between *self* pollen and that produced by other individuals. In most cases, only pollen from sexual partners is able to germinate on the receptive stigmatic surfaces and go on to effect fertilisation. It is known that this recognition is based upon the presence of genetically determined allogens.[72]

Such self recognition has been characterized as a passive, automatic process, being wholly genetically determined. Studies on the communication in plant root systems show that there is much more to the recognition of self in plants than the allogenetic mechanisms employed in avoiding self pollination.[73] Plants are involved in complex and almost constant communication processes with their own organs and other organisms in order to distinguish self from other.

Studies on the grass *Buchloe dactyloides* (Nutt.) Engelm. by Gruntman and Novoplansky have shown that plant roots are able to discriminate between *self* and *not-self*. Importantly, this recognition is based on active, ongoing physiological processes resulting from internal communication within the plant itself. It is not based on a blind allogenetic recognition employed in the self-incompatibility mechanism in plant breeding systems.[74] Although far from completely understood, genetically identical plants have been demonstrated to recognize each other as nonself, suggesting the presence of nongenetic, individually specific signals in the recognition of selfhood and nonselfhood.

Gruntman and Novoplansky speculate that the signalling involved in the identity of plants is "mediated by internal oscillations of hormones such as auxin and cytokinins and/or electricity that are perceived by the roots through the soil."[75] Such oscillating signals "are known to be highly dynamic in time and thus individually unique."[76] It is postulated that these signals are able to be perceived simultaneously inside and outside of the plant. Thus, it is possible that the plant can generate an internal sense of self, through the "resonant amplification of oscillatory signals in the vicinity of other roots of the same plants."[77] That is, organs of the same self actively resonate in the same pattern. Instances of not-self recognition (recognition of different individuals) would occur as "resonant amplification could not occur in roots that are not oscillating with the same rhythm."[78] Recently separated genetically identical individuals would also be expected to oscillate with the same rhythm. The plant synapses, therefore, are likely to be the structures that "determine the integrity of individual plants by allowing their cells and organs to define and detect 'self' and recognize 'not self.'"[79]

This recognition of selfhood is also indicated by studies on soy bean plants by Gersani et al. When soy bean plants were made to share their growth space with other individuals, they substantially increased their root growth.[80] Plants that were given shared resources produced 85 percent more roots than those which did not have a shared growth space. In order to actively and aggressively proliferate their roots in competition for resources, Gersani et al. suggest that these plants must have the ability to perceive and identify the presence of self and nonself roots.

In a complex root network, the ability to differentiate between the roots of oneself and the roots of one's neighbors helps plants avoid the undesirable scenario of some of their roots competing for resources with other parts of their own root network. This would be a wasteful allocation of resources to the roots that could otherwise be used for above-ground functions such as stem growth and flowering. From the premise that natural selection would act against the wasteful use of resources, researchers have predicted that communication processes which recognize self should be widespread among the plant kingdom.[81]

Relating and Interacting Below Ground

Investigations in the rhizosphere also uncover the fact that plants are incredibly interactive. As well as providing water and nutrients, roots can act as conduits of communication between individual plants and other organisms. The rhizosphere is an overcrowded space inhabited by the roots of plants and also by fungi and soil microbes. It is a space in which "root-root and root microbe communications are continuous occurrences in this biologically active soil zone."[82]

Roots are primarily used as communicative tools in the process of sequestering belowground resources such as water, nitrogen, and trace minerals. These resources are fundamental to the survival of photosynthetic plants. An increasing body of evidence is demonstrating that in many cases the method by which roots communicate with other organisms is through the production of root exudates—an umbrella term for an extremely diverse collection of chemical compounds also known as *secondary metabolites*. These communicative root exudates are vital for the flourishing of plant life. Their production commands up to 20 percent of an individual's photosynthate, and they are essential in establishing and maintaining symbiotic, mutually beneficial relationships with soil organisms.[83]

The symbiotic relationship between leguminous plants and *Rhizobium* bacteria is a good example of plants using communication to increase resource uptake, in this case nitrogen availability. Studies across the plant kingdom have shown that the production and secretion of *flavonoids* actually activate the genes of the colonizing *Rhizobium*, which are responsible for producing the root nodule.[84] This communication of information, therefore, directly enables the *Rhizobium* to fix atmospheric nitrogen for host legume plants in nitrogen poor soils. The composition of these flavonoids, and hence, the information content of this root to microbe communication is known to vary greatly between legume species. Each legume species exudes a specific flavonoid composition when communicating with soil microbes. Each has its own "chemical signature," which allows the soil symbionts to distinguish the roots of their symbiotic partners from those of other plant species.[85]

The secretions of the tomato plant *Solanum lycopersicum* L. also allow the creation of mutually beneficial relationships. The provision of information rich root exudates by the tomato is the first step in initiating a root colonization process by the bacterium *Pseudomonas flourescens* Migula.[86] The principal exudates produced by the plant roots, which act as an invitation to establish this symbiotic relationship, are malic and citric acid. This communication through organic acid exudates leads to a relationship in which *P. flourescens* protects the roots of the tomato against parasitic fungi and phytophagous nematodes.

As do humans and other animals, plants actively modify their local environment in order to reduce harmful circumstances and to make it more suitable for growth and reproduction. This behavior is a form of interaction with the local environment, most often demonstrated by plant roots. In the presence of rhizospheric pathogens, plant roots can selectively sense the specific threats and exude antimicrobial and antifungal compounds such as ferulic acid, rosmarinic acid, butanoic acid, and vanillic acid.[87] These chemical signatures are released in a precise response to a specific identified threat.[88] By exuding root secondary metabolites, plants alter the composition and abundance of the soil microflora and reduce the chances of local infection of the roots. Plant roots can also prevent against fungal attack by the production and exudation of defense proteins. Pokeweed, *Phytolacca americana* L., is known to exude defense proteins, which inhibit protein synthesis and have antifungal properties against root rot causing fungus, *Rhizoctonia solani* J.G. Kühn.[89] By removing this pathogen from the rhizosphere, *P. americana* is able to prevent systemic infection, avoid using resources to counter infection and thus increase its fitness.

A further example of active environmental modification is the ability of plants to increase the levels of available phosphorous (P) in soils that are phosphorous poor by the exudation of organic acids and enzymes (acid phosphatases). In lupines which are grown in P-deficient soils, the secretion of these enzymes that help make phosphorous more readily available can increase up to twenty fold.[90]

Allelopathic interaction with other plants is another form of complex rhizospheric interaction. Allelopathy occurs when plants use secondary metabolites secreted from their roots to suppress the growth of neighboring plants. Again, this is a process of modifying the immediate environment in order to increase the chances of growth and reproduction. Over a period of fifty years, there has been a fluctuating acceptance of allelopathy in the scientific community with recent studies suggesting that it has an important role in plant interactions with the environment.[91]

One of the most cited occurrences of allelopathy is in the Mojave Desert shrub community in North America. In an environment where water is a scarce resource, Mahall and Callaway discovered that the common shrub *Larrea tridentata* (DC.) Coville suppresses the growth of neighboring plant *Ambrosia dumosa* (A.Gray) W.W.Payne through the secretion of a readily diffusible inhibitory substance from its root system in order to remove interspecific competition for water.[92]

Allelopathic interactions are also known to occur intensively in crop plants and their neighboring species. Studies show that these interactions are widespread and that they can have a sizeable effect upon growth and reproduction. A common weed of tropical agricultural systems, *Cyperus rotundus* L., can greatly reduce the stem size, leaf size, and overall yield of rice (*Orzyza sativa* L.) through the production of growth-inhibiting organic acids. The presence of *Cyperus* also has the same effect on other important crops such as maize, barley, tomato, and onion.[93]

As well as communicating with microorganisms and beneficially modifying the environment, roots are known to act territorially. In order to manifest this behavior, roots must be involved in continually occurring complex communication, both with the roots of the self and the roots of other individuals. There are many fascinating examples of such communication in root systems. Roots of the clonal herb *Hydrocotyle* have been shown to veer away from competition with the roots of other species.[94] Studies on the desert shrub *Ambrosia dumosa* have shown that individual plants avoid proliferating their root systems if they come into contact with other individuals of their species.[95] Research on the garden pea *Pisum sativum* L. has demonstrated that roots do deliberately avoid contact with other roots on the same plant.[96] The parasite *Triphysaria versicolor* Fisch. & C.A.Mey. has been shown to be able to actively distinguish its own roots from the roots of its host and other species, which is further evidence for the recognition of self and nonself.[97]

Plant Communication for Flourishing

Active communication and interaction between plants and others also occurs above ground. The most obvious occurrences of aboveground communication and interaction occur during plant reproduction and plant-herbivore relationships. In this context, there is mounting evidence from studies of chemical signalling that plants have a high level of awareness of their local environment, and exploit information on their attackers and on their resources.[98] A major way in which plants communicate with the (above ground) world around them is through the airborne emission of volatile organic chemicals (VOC's).

Flowers are known to release a myriad of scented chemicals including fatty-acid derivatives, benzenoids, and terpenoids.[99] Although it seems obvious that flowers emit these scents to attract pollinators, there have been remarkably few scientific studies which have actually demonstrated this communication. One of the few studies to have done so also involves a process of deceit on the part of the flower toward the pollinator in question. The South American orchid *Ophrys sphegodes* Mill. emits simple volatile chemicals, alkenes and alkanes, as well as providing visual cues to elicit the "pseudocopulation" behavior of the *Andrena nigroaenea* Kirby males.[100] These compounds precisely mimic the scent of the sexually receptive female *A. nigroaenea*, and as the male is attracted to the plant for "copulation," pollination is effected. After copulation, *O. sphegodes* flowers emit a compound that is released by nonreceptive female *A. nigroaenea*, which serves to inhibit further copulation, thus directing pollinators to nearby unpollinated flowers.[101]

Plants are known to emit volatile compounds from their leaves to communicate with herbivores in order to deter attack. It has been shown that tree species, which emit high levels of VOC's, are attacked much less by *Coleoptera* beetles than other tree species, which are in the same plant family (Lecythidaceae) but do not emit VOC's.[102] In addition to producing direct defenses, plants may also respond by the airborne emission of volatile organic chemicals to elicit indirect defenses. These VOC's serve as an indirect defense by attracting arthropod carnivores to prey on the attacking herbivorous insects.[103]

As well as providing information to arthropods, the VOC's that are emitted into the atmosphere also provide information about the type of attack, and intriguingly, this information can be used by neighboring plants. Plants have an ability to differentiate between the herbivores that attack them, and so the composition of the chemical blend emitted by damaged plants is different for each type of herbivore. It is also specific to the plant species.[104] A study on the lima bean *Phaeseolus vulgaris* L. has demonstrated the fact that the particular "blend" of volatile organic compounds from plants undergoing herbivory is specific to the type of damage being inflicted.[105] Due to this discrimination between herbivores, the chemical information in VOC's is very specific and can allow neighboring undamaged plants not only to perceive an imminent threat, but also to assess the actual risk faced. This discrimination can allow the undamaged plant to assess the costs of producing its own direct chemical defenses against the likelihood of serious damage occurring through herbivory. After processing the information gleaned from the neighbors, a plant will execute the response which will ensure greatest fitness.[106]

This above ground communication between plants has been much more hotly debated than the communication between roots.[107] Many researchers regard early studies as having been inadequately controlled, and these have been widely discounted. Other contentions lie in the interpretation of results which conclude that plants sending and receiving VOC's may both benefit. Regardless of the early doubts, an increasing number of studies are presenting evidence for above ground communication.[108]

In laboratory experiments, plants can perceive volatile signals in the environment produced by other members of their own species and by other species.[109] Studies have shown that exposure to volatile organic compounds produced as a result of herbivory can directly affect the expression of "defense genes" and often results in the production of metabolites linked to defense, such as terpenoids.[110] However, in the field, only a handful of studies have adequately demonstrated that plants use the information emitted from neighboring plants for the production of defenses. One such example is a field study of the alder, *Alnus glutinosa* L., which is attacked by a herbivorous beetle. In the field, the defoliation of individual trees was found to result in reduced herbivory in neighboring *A. glutinosa* trees. The "protective effect" lessened with distance from the tree undergoing herbivory and from the time since defoliation.[111] This suggests that the damaged trees produced signals that warned of herbivory and which led to the production of antigrazing defenses in their neighbors, enabling increased biological fitness.

Although it is clear that the plants "on the receiving end" are able to perceive this communication, it is widely doubted that these signals are *sent* intentionally by the plant being eaten. Doubts rest mainly on the evolutionary premise that there could be no benefit to those plants sending signals in increasing the competitiveness of neighbors. However, the recent revelation of kin recognition in plants suggests that signals could be emitted in accordance with kin selection theory—that is, the individual signalling plant could intelligently increase its own inclusive fitness by increasing the fitness of related individuals.[112]

Evaluating the Advance

This increasing body of evidence in contemporary plant science is beginning to demonstrate convincingly that plants share many capacities and capabilities with human beings. But it must be remembered that such evidence is at the cutting edge of science. Much of the material is still under scientific debate, and many of the concepts involved in plant intelligence are openly opposed. Struik et al. argue strongly against the use of notions such as intelligence, reasoning, and problem solving as they argue that to use them "surmises extremely complex mechanisms and structures to be present"—which in their view violates the principles of parsimony.[113] Yet, such resistance ignores the fact that extremely complex communicative pathways are now being uncovered in plants and that plants use them to demonstrate complex and intricate behavior. As Brenner et al. state "We are less concerned with names than with the phenomena that have been overlooked in plant science."[114] Although doubts exist—as has been the case at many points in the history of plant sciences—future investigation, especially within plant neurobiology, is likely to only provide further proof of active, intelligent plant life.

Physiologically, the current advance from the perception of plants as inert lumps has been achieved by studying the behavior of individual, free-living plants. These plants are more likely to demonstrate intelligent behavior in order to maximize fitness. Increasingly sophisticated experimental methods and technology has also been of huge importance. Much of the empirical evidence for plant intelligence has been gathered using measurements of plant molecular signalling. Yet, as much of a plant's ecological behavior takes place below ground, technological capabilities are still limiting a full evaluation of plant intelligence.

However, the advances in plant understanding can also be said to have a subtly important philosophical basis. In the terminology of Bashō, "plant scientists have gone to plants to learn about plants."[115] That is, they have conducted close observations of plant life, and instead of *evaluating* this evidence from a *zoocentric* perspective; they have followed Theophrastus in using animal models to consider plants as much as possible on their own terms. Perhaps unconsciously, they are taking a more phenomenological approach to their studies. In the case of Anthony Trewavas, his orientation toward emphasizing the connectivities between plants, humans, and animals, has been adopted explicitly in order to generate a greater respect for the plant kingdom and the wider natural world.[116]

Such a drive toward connection has its own parallels with the treatment of plants in animistic traditions, as well as in Hindu philosophy and Jainism. For the Western mind, the significance of the contemporary plant sciences is that they provide a wealth of empirical evidence for the sentience and intelligence of the plant kingdom. As for these diverse traditions, the ramifications of this empirical evidence for connection, sentience, and intelligence must extend into questions of interspecies ethics.

Importantly, whatever the current scientific debates, the intellectual basis for treatments of plant life as inert, vacant, raw materials is demonstrably false. On this basis, it must be accepted that the continued denial of plant autonomy and the exclusion of plants from human moral consideration is no longer appropriate. This exclusion deliberately treats plants as less than they are. More strongly, such deliberate undervaluing can be considered as a form of intellectual violence.[117] Including plants within human moral consideration is more appropriate than exclusion as it both recognizes and reveals plant sentience. The concluding chapter, therefore, tackles the remaining questions. What shape should human-plant ethics take? How can we move from a stance of exclusion and domination to one of inclusion and care? How can plants be incorporated into dialogical relationships?

Recreating a Place for Flourishing

A fool sees not the same tree that a wise man sees.[1]
The tree which moves some to tears of joy is in the eyes of others only a green thing that stands in the way.[2]
—William Blake

Based upon Erazim Kohák's insight that cultural perceptions are crucial matters in questions of ecology, the preceding chapters have discussed human perceptions of, and relationships with plants, using a wealth of material from a variety of worldviews. My analysis of Western philosophical and religious writings makes it clear that the Western attitude toward plants is zoocentric and hierarchical. A feature of such hierarchies is that they arise in conjunction with the need to justify untrammelled human resource use—the emergence of hierarchy precedes the act of domination.[3] It precedes acts of commodification and ownership. In order to maintain hierarchical ordering, the continuity of life has been ignored in favor of constructing sharp discontinuities between humans, plants, and animals. Shared characteristics such as life and growth have been rejected in order to focus on the gross differences.

The stress on discontinuities between humans and nature is characteristic of Western thought. But along with the usual suspects of mechanism and atomism, we can point to the backgrounding of plants as a key element in human separation from nature.[4] Even within the environmental movement, pioneering studies which have done much to champion the causes of nonhumans (including much animal rights theory) are predicated on the basis that moral consideration should not be given to plants.[5] In the work of Peter Singer, plants are excluded because of the presumption that they lead a "subjectively barren existence."[6] Such processes of exclusion are the grounds for the undervaluation of plants in Western society.

Morally Considerable Photosynthesizers?

Through several chapters, I have shown that many of the criteria signifying moral considerability *can* be located in the plant kingdom. Close observation of plant life-history demonstrates that plants are communicative, relational beings—beings that influence and are influenced by their environment. They also reveal that plants have their own purposes, intricately connected with finding food and producing offspring. Like other living beings, plants attempt to maintain their own integrity in changing environmental conditions. Plants display intelligent behavior in order to maximize both their growth and the production of offspring.

Despite these findings, I do not wish to argue for moral considerability based upon provable criteria. A drawback with approaches that attempt to *prove* moral considerability is that they struggle with the infamous gulf between questions of fact and the impetus for moral action, the "is-ought" gap identified by David Hume. Within the Western ethical framework, the fact that a plant or animal has interests or intrinsic value does not automatically require those intrinsic values to be respected. This is a well recognized gap between a factual description of a being's attributes and the need to subscribe to an ethics that takes these into consideration. As Karen Warren notes, this gap leaves ample room for scepticism.[7]

Instead of trying to prove the existence of criteria, I base the recognition of plants as morally considerable upon the ground of Erazim Kohák's philosophical ecology. Grounded in phenomenology, for Kohák, a description of reality is not a true, definitive description; it is not something that can be ultimately proven. Instead, it must be conceived as a "manner of speaking" about the world because "reality is always what it is and it is vastly more than we can say about it."[8] In this view, different "manners of speaking" are impossible to prove as "true." This is because "Reality *in itself*, abstracted from all lived experience, could have no meaning. Meaning is a relational reality."[9] Truth and meaning require experiencing subjects. In this phenomenological understanding, "manners of speaking" are not objective truths, but nor are they "mere descriptions"; they are "modes of interacting with reality—which render our world meaningful and guide our actions therein."[10]

Because of the effect that "manners of speaking" have on the way we approach the world, the differences between them are differences that matter. Each choice paves the way for different modes of relating with the world—leading to very different ends. In this context, we have a choice between "treating trees as raw materials or treating them with respect."[11] Instead of attempting to prove that one way or another is right, Kohák simply insists that the notion of appropriateness should guide our choice. In an ecological context, if we wish for health and well-being, then appropriate ways of relating to other beings are those that increase connectivity and allow the growth and continued existence of individuals, species, and ecosystems.[12] Purely instrumental relationships clearly do not fulfil the criteria of ecological appropriateness. By removing limits to the actions of human beings, purely instrumental relationships are one of the major drivers of ecologically destructive behavior.[13] Our wholly instrumental relationships with plant life are inappropriate because they are a very significant contributor to the current anthropogenic environmental predicament.

In the *Death of Nature,* Caroline Merchant asserts that the development of the idea that *nature* was passive rendered it freely open to manipulation by Western societies.[14] Here I put forward a similar analysis based on the plant kingdom, bearing in mind that plants underpin life on Earth and form a large part of the biotic and abiotic sphere, which is commonly understood as *nature*. Although there are many complex drivers of environmental degradation, a worldview that regards plants primarily as resource-objects, as materials, is influential. As Plumwood writes, if "nature is a passive field for human endeavour" then it is "totally available for its owners remaking as they see fit."[15] With the knowledge that plants form the basis of natural ecosystems, one of the reasons why nature is a passive field for human endeavor, is that plants themselves have been rendered as passive objects—totally available for unrestrained use by human beings.[16]

This lack of care and respect toward plants has significant environmental effects. The continued alteration and destruction of natural habitats by human beings is one of the major drivers of environmental degradation, species extinction, and global climatic change.[17] Natural habitats are predominantly plant habitats. Natural habitats are populations of plants that exist in relationships of mutual benefit with the birds, fungi, bacteria, reptiles, mammals, humans, and other species. Our ecologically inappropriate behavior toward plants is exemplified by the fact that the rate of habitat clearance by humans is at its historical maximum.[18]

During the last three centuries, twelve million km^2 of forests and woodlands have been cleared; five million km^2 of grasslands have been lost; while cropland areas have grown by twelve million km^2.[19] While the extent of temperate forest vegetation shows signs of recovery, tropical forest destruction proceeds at 130,000 km^2 per year.[20] This assault on plant habitats now directly threatens between 20 to 30 percent of plant species and up to 40 percent of all species with extinction.[21] As a primary driver of global climatic change, it also indirectly threatens biospheric integrity—with an early IPCC report estimating that "about 10 to 30% of the current total anthropogenic emissions of CO_2 are . . . caused by land-use conversion."[22]

Purely instrumental human-plant relationships are ecologically inappropriate, but they are also inappropriate on the basis of their *degree of fit* to the available evidence. Although a purely objective truth is not possible to attain, the most appropriate way of relating should also be the mode of relating that best fits the evidence from the world around us.

Rather than confirming the idea of plants as passive resources, Chapter 7 has demonstrated that the plant sciences contain a wealth of data that indicates the existence and expression of autonomy and intelligence—attributes shared between plants, animals, and human beings. In recognition of these characteristics, treating plants with moral consideration is simply more appropriate than relating to them solely within the old instrumental framework. Perceiving and relating to them as passive resources is outdated and rests ultimately on inadequate observations. Our alternative evidence shows that incorporating plants within the realm of moral considerability is neither fanciful nor misguided. Giving moral consideration to plants is more appropriate than perpetuating their exclusion.

Respectful Relationships

Considering the range of cultural traditions already surveyed shows the existence of numerous paths to moral relationships with plants and numerous examples of what moral relationships with plants actually entail.

Ancient Indian texts recognize an ecological and karmic link between plants, humans, and animals. Plants are recognized as living, sentient beings with their own purposes and goals. Therefore, plants are considered to be within the realm of moral responsibility, and are appropriate recipients of compassion and nonviolent conduct (*ahimsa*). In the Jain tradition, the ethical ideals of compassion and nonviolence are taken to their logical ends. For Jainas, the killing of any sentient being is a violent deed (it acquires negative karma). Killing plants is not the same as killing a human being, but killing plants is still considered to be violent. Therefore, only the killing of the minimum number of plants required for human subsistence is ethically acceptable. In particular, this extends into choices about plants used for food, and the need to avoid wasting plants that have been killed.

Indigenous societies (including pagan societies) share notions of kinship based upon shared Earthly heritage and substance:

> Animals like family to us.
> Earth our mother,
> Eagle our cousin,
> Tree is pumping blood like us.
> We all one.[23]

On the basis that plants clearly demonstrate self-directed growth, the communicative, sentient, intelligent nature of plants is also established. This is the recognition of plants as *other-than-human persons*—a powerful way of incorporating plants within social and moral relationships of care and nurturing Yet, unlike in the animal rights theory of Francione, *persons* are not exempt from use, a fact which has important consequences.[24] With plants as persons, there can be no "substantial outclass of living beings that are morally excluded in order to locate any viable form of eating which allows an ethical basis for human survival."[25] Uncomfortable or not, there is no dualistic separation of personhood and use. Human persons *must* act harmfully toward plant persons in order to live and the necessary harm done to plant and animal persons is accepted, ritualized and celebrated as a fact of being alive.[26] The elegance of this acceptance is that it then acts as one of the principal driving forces behind respectful relationships. Typically, this manifests in the conviction to only harm plant persons when necessary and to encourage the growth of plants where possible.[27]

Talking to Trees

Dialogical engagement helps form the social relationships which are the root of moral consideration and moral action toward plants. In dialogue, "the parties form a unity of conversation, but only through two clearly differentiated voices" and so "dialogue, unlike monologue is multivocal, that is, it is characterised by the presence of two distinct voices."[28] Therefore, to bring about dialogue, the autonomous, communicative prescence of nonhumans needs to be recognized and affirmed.

In the biosphere, dialogue involves plants establishing connections between communicating species, and these connections often allow the blossoming of life, both for the self and for others.[29] A good example of this dialogue is provided by the ecological interactions of the bilberry (or blueberry), *Vaccinium myrtillus* L. Below ground, as with other ericaceous species, bilberry has mutualistic mycorhizal associations with fungal symbionts. Studies of these *Vaccinium* associations with mycorhizal species such as *Pezicula myrtillina* P.Karst and *Phyllosticta pyrolae* (Ehrenb.) Allesch. have demonstrated that the host plant can benefit from increased phosphate and nitrogen uptake, while the fungus benefits by receiving sugars from photosynthesis.[30]

Aboveground, connectivities between bilberry and others center primarily on flowering and fruiting, which are integral to plant dialogue. The bilberry's pendulous flowers are pollinated by bumblebees, honeybees, moths, and syrphid flies.[31] These pollinators are attracted by the maturation of the flower, a signal that nectar and pollen is being produced. Although bilberry can spread vegetatively, pollination by these insects aids sexual reproduction and the maintenance of genetic diversity in the species. This pollination relationship is mutually beneficial, as are relationships that revolve around fruit and seed. Once bilberry fruits have signalled that they are mature, they are eaten by many birds—including grouse, partridge, pheasant, and ptarmigan.[32] The fruits are also one of the main foods of the capercaillie and during the summer also compose a large part of the diet of brown bears in mainland Europe.[33] The dispersal of bilberry seeds by these animals ensures mutual benefits accrue from these cross-species interactions.[34]

In these interactions, we can detect the multiplicity of actors and voices that characterizes dialogical relationships. For a study focussed on human-plant relationships, Bakhtin's model of dialogue makes it clear that human-plant dialogue is only possible where this multiplicity of plant voices is recognized. Therefore, from a human point of view, human-plant dialogues must be based upon allowing plant "voices" to be heard and plant presence to be felt. In this respect, Bashō's advice to avoid imposing the human perspective on plant life must underpin human dialogue with nature.

Animist traditions show that plant "voices" can be transmitted through narratives in which other-than-human persons are featured.[35] Situated stories, songs, and poems can be powerful aids to the recognition of autonomy and personhood in the plant kingdom.[36] Ritual enactment of our kinship relationships with other-than-humans is another powerful way for human persons to lose the "false sense of themselves as superior."[37] It is important, however, that while expressing the human side of dialogue, we also allow others to "speak" for themselves. Otherwise we risk falling back into destructive monologues.

Allowing plant "voices" to be heard entails approaching plants with openness and allowing plants to flourish. Working for the benefit of plants is, therefore, a *direct way* to build more dialogical human-plant relationships, which ultimately result in moral consideration and action. As Deborah Rose explains, "A dialogical approach to connection impels one to work to realize the well-being of others. . . . The path to connection, therefore, does not seek connection, but rather seeks to enable the flourishing of others."[38]

In the context of an anthropogenic ecological crisis, dialogical relationships can not remain theoretical formulations; they must become direct action.[39] Human dialogue with plants should both recognize the other-than-human person *and* strive to introduce reparations that both acknowledge past violence and aim to lessen future violence. It may be the case that "the first thing a philosopher says to a tree is *sorry*," but apologies to the plant kingdom need to go further. Respectful, moral relationships with plants need to be manifested in our behavior.[40]

Colliding Agendas—Mitigating Harm to Plants

Working toward the flourishing of plants as individuals and as aggregates such as species and communities, represents an opportunity for biospheric repair. Yet, here there is an unavoidable encounter between competing interests, as Chris Cuomo says, "some flourishing must always be sacrificed for the flourishing of others."[41] However, dialogical relationships involving humans recognize the perspective of the other and are aware of when the other been harmed. They are empathetic in the sense of perceiving "the other as being another center of orientation in a common spatial world."[42] In an ecological context, we must balance the recognition of harm done to individuals with that done to entities such as species and communities. It may be more ecologically appropriate to maintain diversity and stability in communities, but this does not preclude the recognition of individual harm.[43] An awareness of the capacity for individual flourishing to be violated helps sustain mutually beneficial cross-kingdom connections, thus preventing the greater violence of dislocation and disconnection.[44]

Seeking to enable the well-being of individual plants, plant species, and communities is also a considerable challenge because the interests of human beings repeatedly impact upon the well-being of plants.[45] In this area of contestation, working toward the well-being of plants firstly demands limits on human claims. Ecological reality dictates that humans must use plants to sustain human life. However, if we commit to working for their well-being, we must first recognize that often our use of plants violates the purposes of plant life. Often, our use of plants for food, for medicines or fuel requires committing harm to or killing aware, intelligent, and perceptive beings that seek to live and thrive in the same way as other living beings.

Perhaps the first way in which this can be lessened is by *examining and reforming our use of plants*. As exemplified by the Jain and Indigenous traditions, taking plant welfare into consideration must force us to examine our individual consumption of the most ubiquitous plant products such as food, paper, wood, and medicines. The ubiquity of plant usage renders it impractical for anyone to construct an exhaustive list of instances where harm to plants may be mitigated. However, there are several key areas where an ethic of dialogical respect can begin to focus. These overlap and interpenetrate, but three broad areas can easily be identified.

The first of these is lessening the wastage of plant lives—that is, treating plant lives as nothing. Wasting plant products, particularly paper and food, drives unnecessary harm to plants. In the United Kingdom, recent studies have suggested that Britain wastes up to one third of all food fit for human consumption.[46] A large proportion of this is plant based. Paper from timber is also wasted to the same degree, with 1.3 million tons of waste paper per year collected from households in England alone.[47] While the figures relate to Britain, similar wastage of plant lives occurs across Western society, and this typifies our instrumental relationship with plants.

Although not as simple to quantify as wastage, the sheer (predominantly Western) overconsumption of plant products, both individually and globally, is another identifiable threat to plant well-being.[48] As has already been stated, on a global scale, over the last three hundred years, twelve million km^2 of forests and woodlands have been felled, and five million km^2 of grasslands have been converted to agriculture.[49] This loss of natural plant habitat is driven by human overconsumption of agricultural plants for food, wood, fuel, and medicines. Much of this stems from Westernized nations with instrumental human-plant relationships. The conversion of autonomous plant habitats into agricultural areas intended to satisfy human purposes threatens the integrity of plant species, ecosystems, and the biosphere as a whole. In Europe alone, over 60 percent of the available land has been converted to farmland for human beings. This is set to bring about a sharp decline in plant diversity.[50]

Western overconsumption is significant for it is one of the drivers of habitat and species loss. Globally, it is estimated that natural habitat loss threatens between 20 to 30 percent of plant species with extinction.[51] In some groups such as the cycads, (the oldest seed-bearing plants on the planet) over 50 percent of the species are threatened with extinction.[52] According to recent studies, approximately fifteen thousand species of medicinal plants are threatened in the wild, as a result of overharvesting, land conversion, and habitat loss.[53] Not only is this loss an enacted violence upon the species in question, but these extinctions have serious consequences for ecosystem functioning. The interactions between plants are vital for maintaining the composition and integrity of ecosystems.[54] The loss of plant species can undermine ecosystem stability. It can also undermine both resource availability and habitat structure, which in turn weakens the ability of ecosystems to respond to environmental changes, such as climatic change.[55]

As the case of medicinal plants exemplifies, when humans use plants, there may be direct conflict between human wants and plant needs. This conflict is felt on all levels—from the individual through the species, up to habitats and ecosystems. At the same time, it must be remembered that not all harm to individuals is ecologically harmful. The bear that eats bilberries may kill some of the individual plants, but at the same time, it spreads the seeds. Consumption can lead to (re)production. However, overconsumption is symptomatic of human-plant relationships that revolve around instrumentalism.[56] The drivers of overconsumption are complex and intricately linked, but include human overpopulation, unequal distribution of wealth, greed, urbanization, and industrialization.[57] Although there are many contributing factors, overconsumption is influenced by the fact that in instrumental relationships with plants, there is no inherent moral limit to human use behavior.

A third very significant driver of harm to individual plants, plant species, and plant habitats is the unnecessary, unthinking use of plants.[58] Perhaps the most prominent of these is the use of plants to feed massive numbers of animals for the world's wealthiest nations to consume. Recent estimates suggest that humankind farms and eats over thirty billion animals each year.[59] In a plant context, this livestock rearing is important because it accounts for more than 65 percent of the total global agricultural area.[60] It also accounts for large volumes of grains and soya beans which are used as feed. In 2002, approximately 670 million tons of grains were fed to livestock, roughly a third of the global harvest.[61] They were also fed 350 million tons of protein-rich products such as soya and bran.[62]

The areas cleared to rear animals and feed them on such a huge scale are natural plant habitats such as tropical forests, savannahs, and grasslands.[63] The rearing of livestock on such large scales is one of the major drivers of habitat loss. Basing diets on meat consumption excessively inflates the area of land that is put under human cultivation. Reducing the amount of consumed meat is a direct way of reducing harm done to plants, animals, *and* human beings. Not least because this large industry is also responsible for generating 18 percent of global carbon emissions—which to provide an idea of scale, is more than all forms of transport combined.[64]

In the desire to reduce carbon emissions from oil-based transport, the world's wealthiest nations are also beginning to use plants to feed their cars. The rise in the use of biofuels has the potential to be another significant contributor to the unthinking overconsumption of both individual plants and plant habitats. Current global production of biofuels is estimated at 2.8 million tons per year, but if biofuels were to replace just 20 percent of our petrochemical demands by 2050, 276.7 million tons per year would need to be produced.[65] This hundredfold rise in the production of the four major biofuel crops[66] also has the potential to lead to the loss of natural plant habitats.[67] A recent study by Pin Koh suggests that using "soybean based biodiesel production to meet future global biodiesel demand would likely result in the highest amount of habitat loss (76.4–114.2 million ha) compared with alternative scenarios of sunflower seed (56.0–61.1 million ha), rapeseed (25.9–34.9 million ha), and oil palm based (0.4–5.4 million ha) biodiesel production."[68] The use of tropical crops such as soybean and oil palm poses a particularly high risk to plant biodiversity as they are typically grown in geographic regions that contain some of the world's biodiversity hotspots.[69]

The irony is that while biofuels are touted as a means of reducing carbon emissions in the fight against global climate change, the most recent studies have demonstrated that the loss of natural habitat required for biofuel production actually increases greenhouse gas emissions.[70] When carbon emissions from land-use change are factored in, in the United States the use of corn-based biodiesel nearly doubles the emission of greenhouse gases over thirty years.[71] The use of food to drive cars also pushes up grain prices, putting poorer nations at risk of malnutrition and increases the pressures on natural plant habitats in developing countries.[72]

Again, the link is evident between lessening the harm done to individual plants, plant species. and plant ecosystems and reducing the harm done ultimately to human beings. We need to fundamentally examine our uses of plants and decide which are necessary and which are not. After fulfilling our basic needs, the needs of plants also need to be recognized. An awareness that there are other subjects and purposes in the biosphere demands limits to human activity. It demands that humans only violate these needs and purposes where necessary, either for the satisfaction of human needs or for the maintenance of biodiversity. Ultimately it should demand the cessation of the wasteful and unthinking use of plant individuals, species, and communities.

Reducing harm is the first step. Because of the extent of the biodiversity crisis and the incipient march of plant habitat loss, we must also find ways to make the space necessary for plants to thrive and reproduce their kind. In fact, this should be recognized as one of the priorities for humankind in the twenty-first century. Practically, leaving space for plants entails expanding protected area networks on a local, regional, and national level.[73] Protecting greater areas of plant habitat from conversion to human use will help to ensure the continuation of wild living plant species and communities.[74]

Conserving plants for their own sake, as well as for the needs of others, would help bolster conservation efforts for marginal, uneconomic plant species. Increasing the area of plant habitat that is unavailable for transformation to human ends will be a practical step toward maintaining biodiversity and mitigating climatic change. Such preservation is also essential "in order for there to be a nature with which to have a relationship."[75] Here there is a distinct connection between the preservation of more natural plant habitats, and the mitigation of violence to plants, predominantly agricultural plants. Reducing the killing of crop plants lessens the area of agricultural land that human beings need to farm. This would reduce real pressures on plant habitats around the world.

Putting Plants First In Restoration

If we wish to engage in dialogue with nature, our efforts need to be active as well as passive; they need to be restorative as well as conservative. Moral relationships with plants need not be restricted entirely to the negation of human claims and the mitigation of harm. As Hettinger says, "leaving much of nature on the planet alone is an absolutely central part of any adequate environmental ethic," but following John Visvader, we also need to "imagine giving more to the world around us than the gift of our mere absence."[76] As contemporary animists demonstrate, caring for the well-being of plants requires social interaction, human presence, and activity.

The restoration of plant habitats offers a way to divert human industry into working for the well-being of plants. Globally there are thousands of restoration projects on a variety of different scales—from the restoration of a local woodland, or a pagan grove, to the billion dollar project to restore the Florida Everglades.[77] The majority of restoration projects involve recreating plant habitats on damaged lands, such as brownfield sites or on previously cultivated or grazed areas. On such sites, restoration is often the only appropriate human response to ecological damage. The success of projects such as Trees for Life's restoration of Scotland's Caledonian forest suggests that restoration is becoming an increasingly popular way of engaging with nature. In many ways, restoration is a way of *enacting* our knowledge of plant personhood and human-plant kinship. It releases us from the theoretical ethical domain and can point us in the direction of ritual—of embodied, performed activity which Grimes argues is necessary for humans to develop the humility and gratitude that will give us all a chance of surviving to the third millennium.[78]

However, the process of restoration is not without its criticisms. One of the most prominent critics of restoration is Eric Katz, whose objections to the restoration of natural habitats are founded on the claim that restored environments are not natural.[79] In this argument, anything achieved through the work of humanity cannot be natural. Therefore, Katz asserts that all restoration projects "involve the manipulation and domination of natural areas. All of these projects involve the creation of artifactual realities, the imposition of anthropocentric interests on the processes and objects of value."[80] However, as Andrew Light contests, the notion that the fruits of human work cannot be natural itself rests on a radical ontological dualism between human beings and nature.[81] This dualism is at the heart of our environmental predicament, and rather than being a weakness of restorative action, it is one of the strengths of restoration that it works to overcome this ecological schism. For William Jordan III, a fierce proponent of restorative action, restoration is not the imposition of human interests on nature, but is "a way of repaying our debt to nature."[82] Restoration of habitats is a way of giving something back to the natural world. It is primarily an active, performed apology for centuries of domination because its aims are fundamentally "to let alone" the beings who are being restored to their habitats.[83] With this motivation, restoration actively gives space for the lives of plant species.

The aim of restoration is different from traditional gardening, forestry, or agriculture. The aim is not to bring plants under human control and deny their subjectivity, but rather to reclaim habitat in which other species can live. While even the most benign agricultural activity involves "simplifying an ecosystem in order to exploit it more effectively for some human end, restoration does the opposite, recomplicating the system in order to set it free, to turn it back into or over to itself."[86] Jordan has argued that this restorative process is primarily done "with a studied indifference to human interests."[85] In his ethos of restoration, this is not a divorce of humans from the natural world, but the avoidance of imposing short-sighted, short-term human "interests" of domination and control on plant habitats.

Human beings do of course have a very real interest in diverse and healthy ecological communities. The continued restoration of the Wagait floodplains in northern Australia is allowing the native Mak Mak people to continue to gather, hunt, eat, and live in their homeland. In this instance, as in many projects, the restoration of plant communities involves the harm of individual plants. For the Mak Mak, the restoration of their wetlands involves the removal of huge numbers of the invasive *Mimosa pigra* L.[86] As Harvey's animist manifesto makes clear, in some instances respect does not rule out the possibility of death for some and life for others.[87]

One of the most positive aspects of restoring plant habitats is its capacity to arrest domination, both mentally and physically. Restoration is an active and practical method for putting plant, mammal, bird, insect, and fungal interests on a par with (and in many projects often before) those of human beings. The active nature of restoration is significant because it allows dialogue with nature, and human plant-ethics to be more than theoretical. It allows human beings to directly replace instrumental relationships with social relationships of care and solidarity by pursuing ecologically beneficial work. It allows the grounding of theoretical ecological relationships in ecological reality and is a way of practically restoring the overall human relationship with nature.[88]

The fact that humans may work to greatly influence the composition, situation, and scope of plant habitats during restoration is not a weakness (as identified by Katz), but is actually a great strength. The influence that we exert in restoring plant habitats "forces us to become aware of ourselves as ecologically effective inhabitants of a world inhabited by others," a process which restorationists often find deeply pleasurable, and which restores humans to a more appropriate understanding of themselves as relational beings.[89] Participating closely in the difficult task of recreating (as closely as possible) wild and free-living ecosystems makes human participants aware of the fragility and intricacy of plant ecosystems that repeatedly come under negative human influence.

Working for the benefit of other-than-human persons allows the reclamation of positive human ecological influence on the natural world. As restored habitats often necessitate continued human attention, restoration creates real relationships of care between humans and plants. These dialogue-based relationships recognize the plurality of voices in the natural world, and are a work in progress toward dissolving the human-nature dualism that is at the heart of our ecological predicament.[90] Thus, the necessity of human engagement with other subjective beings during restoration is one of its key strengths. As Deborah Rose powerfully states, "subject-subject encounter is an ecological process that undermines the whole basis of hegemonic anthropocentrism."[91]

Although it may involve harm to some individuals, direct action to reestablish plant habitats does not vitiate the autonomy of our plant kin. This is because in effect, dialogically oriented restorative action is a collaboration between human beings and plants. This collaboration is based upon humans putting plant needs at the heart of our action. Rather than exerting domination, "restoration can allow nature to engage in its own autonomous restitution."[92] Thus, working closely in collaboration with plants on a restoration project can allow many Westerners the opportunity to directly encounter the autonomous qualities that plants possess.

Although human beings may influence the situation of sessile plants during restoration, restorative activity does not negate the fact that plants are continuously behaving autonomously and intelligently. Regardless of how they come to be fixed in the ground, once rooted, plants begin to grow, perceive, communicate, and plastically alter their forms—making reasoned decisions as they do so. In both a primary forest habitat, and a restored forest, the vegetation will be abuzz with communication and mental activity.

The seed of an understanding that plants are active, self-directed, even intelligent beings can be sown by science, but it must be realized through working closely with plants in collaborative projects of mutual benefit. Working closely with individual plant persons also has the potential to shift the view of nature as an organic, homogenized whole—which by blanking individual personalities contributes to the backgrounding of nature.[93] The earned, practical recognition of plants as persons releases us from the dichotomy of regarding nature either as a combination of *processes* or *things*.[94] Instead, it puts forward the view that nature is a communion of subjective, collaborative beings that organize and experience their own lives.[95]

NOTES

A Philosophical Botany

1. Kohák 1993, 383.
2. See Plumwood 1993 and Plumwood 2002.
3. See Plumwood 2002, 9. Karen Warren regards Western worldviews as sanctioning a "logic of domination"; see Warren 2000, 46–47.
4. UNEP 2007, Foreword xvi.
5. For an engaging discussion of different nature concepts and their politics, see Soper 1995. For a response to the deconstruction of the concept of nature, see Soulé and Lease 1995.
6. Harvey 1996.
7. Mathews 1999, 120.
8. Wolch and Emel 1998, Introduction xv.
9. Regan 1983; Singer 1995; and Francione 2008.
10. See Bekoff and Goodhall 2002 and Bekoff 2003.
11. Ingold 1994; Wolch and Emel 1998; Emel, Wilbert, and Wolch 2002; Bekoff and Goodhall 2002; and Bekoff 2003.
12. As plants and nature are practically synonymous, it is perhaps appropriate then that the very first use of the term *nature* in Western literature was used in connection with the plant world. The Greek *physis,* translated into Latin as *natura,* can be traced back to Homer and a description of a plant. *Odyssey* 10.302–303.
13. Trewavas 2003.
14. For an account of the invisible and cryptic prokaryotic biodiversity of Earth, see Nee 2004.
15. By focussing solely on plants, I do not wish to set up dichotomies and exclusions between living and nonliving constituents of the natural world. I

focus here on plants because of their fundamental role in sustaining life on Earth, and because the question of human-plant relationships has received little attention in contemporary nature debates.
16. Kohák 1993, 375.
17. Kohák 1993, 380–381.
18. Kohák 1993, 381.
19. Mathews 1994.
20. Deutsch 1986, 294.
21. Bakhtin 1984, 293.
22. Hallé 2002, 17.
23. Wandersee and Schussler 1999, 84–86.
24. Wandersee and Schussler 2001, 2–9.
25. Hallé 2002, 17.
26. Wandersee and Schussler 2001, 5–6.
27. Ibid.
28. Wandersee and Clary 2006, 2.
29. Ram-Prasad 2007, 9.
30. Plumwood 1993.
31. Warren 2000, 47.
32. Biehl 1997, 75.
33. Warren 2000, 46.
34. Ram-Prasad, 9.
35. Coleman 2006, 535.
36. Warren 2000, 46.
37. This alternative Western heritage is discussed in Chapter 5.
38. Mechanistic and atomistic approaches to nature have been analyzed by Leiss 1972; Merchant 1982; and Mathews 1991.
39. In Plato's *Republic*, the construction of a hierarchy in human society was based upon the need to reduce violence in society.
40. A heterarchy is an organizational structure resembling a network rather than a top-down chain of command. Margulis and Sagan 2000 call a nonhierarchical relationship between life forms a *holarchy*.
41. Bakhtin 1984, 293.
42. Kohák 1993; Thompson 2007, 386, regards empathy as "a unique form of intentionality in which we are directed toward the other's experience."
43. See Chapters 5 and 6.
44. For Chris Cuomo, *flourishing* beings are those capable of adapting to change, maintaining their physical integrity, and displaying good or ill health. Those beings that flourish are connected in communities that facilitate this flourishing. Thus, flourishing beings can be harmed by restricting their adaptation, violating their physical integrity, and by causing their ill health. Cuomo 1998, 71–74.

45. Harris 1991, 58–59.
46. As Levinas astutely observed, the recognition of sameness and difference is needed for the construction of relationships.
47. Rose 1999.
48. This understanding has much in common with Maturana and Varela's (1980) notion of *autopoiesis*.
49. Brennan 1988, 138.
50. Plumwood 1999, 200–201.
51. Callicott 1997, 2.
52. Harvey 2005a, 172.
53. Callicott 1997, 10.
54. Bauman 1993; Kohák 1993; and Heyd 2005.
55. Plumwood 2002, 9.
56. Rose 1999, 184.
57. Rose 2006, 77.
58. Bakhtin 1981, 1984.
59. Baxter and Montgomery 1996, 24.
60. Bakhtin 1984.

Chapter 1. The Roots of Disregard

1. Plato *Timaeus* 90a. All versions of Plato's works are taken from Cooper (1997).
2. Mourelatos 1993, 4–5.
3. Carone 2003, 67–80. Also see Egerton 2001, 93–97.
4. *De Anima* 411a7–8, *DK* 13A10 and Mourelatos 1993, 4. Quotes from *De Anima* are taken from Lawson-Tancred 1986. Quotes from other works of Aristotle are from Barnes 1984.
5. *Physics Fr.* 14 (21).
6. This is also a feature of the biblical account of plant creation, see Chapter 3.
7. *Physics Fr.* 78 (107).
8. Wright 1981, 235.
9. *Karthamoi Fr.* 122 (136).
10. *Karthamoi, Fr.* 127 (140).
11. *Theogony* 2.116–138.
12. Mourelatos 1993, 7.
13. *Stanford Encyclopaedia of Philosophy*, www.plato.stanford.edu/entries/presocratics.
14. Plumwood 1993, 72.
15. Plumwood 1993, 81.
16. Plumwood 1993, 82.

17. Vlastos 1973, 34.
18. Vlastos 1973, 154.
19. Plumwood 1993, 82–86.
20. *Timaeus* 30.
21. Plumwood 1993, 84–85; *Timaeus* 50.
22. In *Laws* 966b, Plato describes the state of mind of a slave as one lacking in reason. And as Plumwood (1993, 87) notes "Appetite or desire, the lowest section of the soul, is clearly and constantly identified with wild animals and with 'children and women and slaves' *Republic* 431C." In the same passage, Plato regards the rational mind to be the preserve of the most educated, excellent people—an elite minority primarily composed of men.
23. These three levels of the soul are identified in the *Republic* 439e–441c and are explicitly hierarchically ordered with reference to the social hierarchy in the polis.
24. Plato also regards reason as the natural ruler of baser nature 441e. In the *Timaeus*, the creator god decrees that intelligent beings are superior to non-intelligent beings (30b).
25. Ram-Prasad 2007, 9.
26. In the Greek version, the word used is *zoion*, which of course is the root word of zoology, the study of animals. However, *zoe* also means life and *ion* denotes location. Therefore *zoion* can also be translated as a "place of life" or more simply a living being. Perhaps it says more about the translator's view of plants that this is translated as "animal."
27. *Timaeus* 77abc.
28. Carone 2003, 72.
29. See Vlastos 1968.
30. Vlastos 1973, 160.
31. *Timaeus* 90a.
32. Plumwood 1993, 71.
33. *De Anima* 412^a.
34. *De Anima* 413^a.
35. *De Anima* 416^a.
36. Ibid.
37. *De Anima* 414^a.
38. Carone 2003, 74.
39. *De Anima* 413^{ab}.
40. *De Anima* 414^{ab}.
41. *Nichomachean Ethics* 1102^b.
42. See Chapter 2.
43. See Lovejoy 1936 and Mayr 1969.
44. Carone 2003, 74.
45. *Nichomachean Ethics* 1102^b.

46. Ibid.
47. *De Anima* 430^a.
48. See Plumwood 2002.
49. *Politics* 1256^b.
50. *Politics* 1254^a.
51. Plumwood 1999.
52. Vlastos 1973, 160.
53. These points add weight to the claim that such backgrounding is a deliberate exclusion.
54. *De Anima* 424^{ab}. My emphasis.
55. *De Anima* 414^b.
56. Plumwood (1993) illustrates a similar backgrounding process in Plato's treatment of women.
57. *De Anima* 412^{ab}.
58. *Parts of Animals* 655^b.
59. Of course while it is true that plants do not digest food in the same way as animals, plant roots do not passively take in food from their environment. They exercise control of the water entering the root by means of the waterproof Casparian Strip, and they can control the uptake of CO_2 through their stomata, and increase their light resources by moving away from shaded areas. Through the autotrophic process of photosynthesis, plants produce waste, like animals—waste in the form of oxygen that is vital for our human lives. Of course Aristotle was unable to observe this active excretion.
60. Sarton 1952, 547.
61. Ibid.
62. *HP* 1.1.9–11.
63. *Metaphysics* 8.3.
64. *HP* 1.1.2–3.
65. *HP* 1.4.5.
66. *HP* 1.1 3–4.
67. *HP* 1.2.3–5.
68. Certainly, an explicit hierarchy of the natural world does not appear in the surviving work of Theophrastus. Furthermore, the natural existence of such hierarchies is contradicted by Theophrastus's anatomical work and his views on sense perception.
69. *Metaphysics* 8.2.
70. *HP* 1.2.3–5.
71. *HP* 1.2.1–3.
72. *HP* 1.2.5–6.
73. *HP* 1.2.3–5. "However since it is by the help of the better known that we must pursue the unknown, and better known are the things larger and

plainer to our senses, it is clear that it is right to speak of these things in the way indicated [i.e., the comparison of plant parts with animal parts]."
74. It appears that Theophrastus recognized the significance of these claims, because he himself draws attention to a couple of remarkable passages.
75. *CP* 1.16.1.
76. *CP* 1.16.2 "Compare cultivated to wild, tended to untended, better tended to tended worse and in practically every case the former has smaller stones, is more fluid and diverts the food more to the pericarpion; it moreover ripens the juice to the point where this is adjusted to man's requirements."
77. *CP* 2.16.6.
78. *CP* 1.16.11.
79. *CP* 1.16.6.
80. *HP* 1.1.
81. Browning Cole 1992, 53–55.
82. Priscian, *On Theophrastus on Sense-Perception*, 1.1 and 20.9.
83. *HP* 1.7.1 My emphasis.
84. See also *CP* 1.12.8; "Plants, both young and old, have in all their parts an impulse to grow."
85. Egerton 2001.
86. Theophrastus recorded this practice from Babylonia.
87. *CP* 2.3.3.
88. *HP* 2.7.1.
89. *CP* 2.3.7.
90. *CP* 2.3.7–4.1 and 9.1.
91. *CP* 1.10.5.
92. *Metaphysics* 2.3.
93. *CP* 2.11.1.
94. *CP* 2.11.3. My emphasis.
95. *HP* 2.7.6.
96. *CP* 1.16.12.
97. See Kohák 1997.
98. Hallé 2002.

Chapter 2. Dogma and Divination

1. Leclerc 1753, 109.
2. H. Boerhaave, quoted by Linnaeus. Quoted in Morton 1982, 232.
3. For detailed studies of the impact of these thinkers see Weiss 1972; Merchant 1982; Mathews 1991; and Plumwood 1993.
4. Healy 1999, 387.
5. Pliny is often accused of copying chunks of earlier works. See Morton 1982.

6. *Natural History (NH)* 16.72.
7. *NH* 16.58.
8. *NH* 17.21–24.
9. *NH* 17.2.
10. Cicero *Concerning the Nature of the Gods*, quoted in Passmore 1980, 14.
11. *NH* 18.1.
12. *NH* 16.24.
13. *NH* 22.7.
14. Leiss 1972; Merchant 1982; and Mathews 1991.
15. Plumwood 1993.
16. Morton 1982.
17. This bias persisted well into the sixteenth century; for even when the first botanical gardens were established at Pisa and Padua, they were for the study of medical plants.
18. Morton (1982) notes that in the Middle Ages the theory of knowledge of St. Augustine was dominant. It held that knowledge and sciences were a reflection of the divine in the human mind. Harking back to Plato, it rejected the Peripatetic view that knowledge derived from the application of thought to experience. Its dominance turned minds away from observation and experiment to reliance on illumination by authority, either by that of the Church or through authoritative texts. In this way, for botany, people disregarded their own experience of plants (or preferred not to look at them at all) and just regurgitate the work of Dioscorides and Pliny.
19. Indeed, for Dioscorides, his text was also extremely influential in the Arab world, through the studies and translations of the Nestorian scholars first in Persia and then in Baghdad. From accurate and beautifully illustrated translations of original manuscripts, the work of Dioscorides became the basis for the Arab pharmacopoeia. See Morton 1982, 87–88.
20. "Thus the life of plants is said to consist in nourishment and generation; the life of animals in sensation and movement; and the life of men in their understanding and acting according to reason." *ST* II–II, Q. 179, A. 1.
21. *Summa Theologica (ST)* II–II, Q.179, A. 1.
22. "Dumb animals and plants are devoid of the life of reason whereby to set themselves in motion; they are moved, as it were by another, by a kind of natural impulse, a sign of which is that they are naturally enslaved and accommodated to the uses of others." *ST* II–II, Q. 179, Repl. Obj. 2.
23. Mayr 1969.
24. Arber 1950, 29.
25. Cesalpino, *De Plantis Libris* quoted in Sachs 1890, 43.
26. Sachs 1890, 43.
27. Sachs 1890, 29–39. Aristotle, *Movements of Animals* 703a.
28. Bacon 1670, 89.

29. Webster 1966.
30. "The corn typha, and spelt, are changed into wheat, and wheat into them for this may be done, if you take them being of a thorough ripeness, and weaned them, and then plant them, but this will not so prove the first nor the second year, but you must expect the proof of it in the third year, as Theophrastus shows. Pliny writes, that the Corn Siligo is changed into Wheat the second year." Della Porta 1658, 62.
31. Della Porta 1658, 8.
32. Ibid.
33. "By reason of the hidden and secret properties of things, there is in all kinds of creatures a certain compassion, as I may call it, which the Greeks call Sympathy and Antipathy. But we term it more familiarly, their consent, and their disagreement. For some things are joined together as it were in a mutual league, and some other things are at variance and discord among themselves. Or they have something in them which is a terror and destruction to each other, whereof there can be rendered no probable reason." Della Porta 1658, 8.
34. Della Porta 1658, 8–9.
35. Della Porta 1658, 27.
36. Webster 1966.
37. Ibid.
38. Mathews 1991, 32.
39. Bacon 1670, 125.
40. Bacon 1670, 126.
41. See Chapter 2.
42. Bacon 1670, 126.
43. Ibid.
44. Bacon 1670, 496 and Webster 1966, 10–11.
45. Webster 1966, 11.
46. Arber 1950, 34.
47. Ibid.
48. *Isagoge Phytoscopica*, quoted in Morton 1982, 169.
49. Morton 1982, 169.
50. Arber 1950, 33.
51. Weiss 1972; Merchant 1982; Mathews 1991; Plumwood 1993; and Szerszynski 2005.
52. Haldane and Ross 1967, 289.
53. Haldane and Ross 1967, 106.
54. Garber 2002, 191.
55. Letter to Debeaune, quoted in Garber 2002, 191.
56. Mathews 1991, 17.

57. Haldane and Ross 1967, 106.
58. The denial of sensation and purpose in animals stands in contrast to the treatment of animals found in the work of Aristotle.
59. As many commentators make clear, Descartes view of mind and matter was sharply dualistic. See Plumwood 1993.
60. Garber 2002, 200–204.
61. See Mathews 1991, 16–18.
62. Morton 1982, 234.
63. Morton 1982, 235.
64. Locke, *Essay Concerning Human Understanding* 2.27.5.
65. Locke, *Essay Concerning Human Understanding* 2.27.12.
66. Locke, *Second Treatise of Government* 5.36.
67. Morton 1982, 210.
68. Ray 1798, 28.
69. Ray 1798, 59.
70. "There is no greater, at least no more palpable and convincing argument of the existence of a Deity, than the admirable art and wisdom that discovers itself in the make and constitution, the order and disposition, the ends and the uses of all the parts and members of this stately fabric of heaven and earth." Ray 1798, 32.
71. Ray 1798, 44.
72. On the subject of the vegetable soul, Ray notes that "I make use of this division to comply with the common and received opinion and for easier comprehension and memory; though I do not think it agreeable to philosophic verity and accuracy, but do rather incline to the atomic hypothesis." Ray 1798, 44.
73. Ray 1798, 93.
74. Ray 1798. 94.
75. This term derived from Cudworth strangely portends the recognition that phenotypic plasticity is the basis for intelligent behavior in plants. See Chapter 7.
76. Ray *Historia Plantarum*, quoted in Webster 1966, 20.
77. Ray 1798, 101.
78. Ray 1798, 56.
79. Linnaeus 1735, 105.
80. Linnaeus 1735, 104.
81. "Since science consists in grouping together of like and the distinction of unlike things, and since this amounts to the divisions into genera and species, that is, into classes based upon characters which describe the fundamental nature of the things classified, I have tried to do this in my general history of plants, and what progress my limited talents have been able to

make in this field, I bring forward for the benefit of all. This rational method of proceeding was indeed indicated by Theophrastus amongst the ancients." Cesalpino, quoted in Morton 1982, 135.
82. Leclerc 1753, 109.
83. Plumwood 1999, 193.

Chapter 3. Passive Plants in Christian Traditions

1. Genesis 6:17–19. All biblical quotes are taken from English Standard Version Bible. Online at www.biblegateway.com.
2. White 1967.
3. For an overview of the various interpretations of the biblical attitude to nature, see Callicott 1997.
4. Barr 1972, quoted in Callicott 1997, 15.
5. While ecotheologians have generally accepted that Christianity has neglected nature, critics of White doubt the extent of the "cultural authority" of Christianity and cite other reasons for the Western world's destructive ecological behavior. See Nash 1991; Nash 1989, 87–112; and Berry 2006.
6. Page 1992, 35.
7. Bauckham 2006, 33.
8. Moltmann 1985, 31.
9. Niebuhr 2006, 87.
10. Callicott 1997.
11. Ibid.
12. Hayden, in Callicott 1992, Foreword xxii.
13. Attfield 2003, 98.
14. Katz 1994, 56.
15. Moltmann 1985.
16. Moltmann 1985, 29.
17. Habel 2000, 37.
18. The counter argument, very kindly provided by Sally Alsford, is based upon the idea that the Genesis text is *etiological* rather than *ideological*—seeking only to explain the causes of human reliance upon the natural world for food, rather than establish a hierarchy of ontological difference and domination. As part of this argument, it could be claimed that hierarchy is being "read back" into the Genesis text and that the establishment of hierarchy was not being consciously addressed at the time of its composition. This is an argument that deserves recognition, and it may indeed be partly true. The fact that plants are created before animals is a likely expression of this relationship, which as Callicott (1997) argues, roughly parallels contempo-

rary evolutionary understanding. However it must be clear that an etiological explanation fails to explain and does not remove the clear backgrounding of plants as passive and nonliving beings—a portrayal that as in Greek philosophy exists in conjunction with human claims on plants as natural resources. Plants could be etiologically designated as human food without their reduction to inanimate, insensate beings incapable of being harmed. Indeed, the Genesis text cannot be purely etiological because such an interpretation cannot explain the portrayal of plants as beings that lack the breath of God, the element that makes human beings closest to God, the source of value. Rather than etiological, it is clear that this assignation of levels of value and power places the passages of the Genesis text that deal with the natural world within the realm of the political.
19. This description of creation was assembled by Jewish priests who had been exiled after the destruction of Jerusalem in 587 BCE and forms a part of the Bible known as the Priestly Document.
20. Moltmann 1985, 14.
21. Moltmann 1985, 9–10.
22. Moltmann 1985, 14.
23. Genesis 1:11–13.
24. Genesis 1:20–21.
25. Genesis 1:24.
26. Hillel 2006. See Chapter 6 for parallels with ancient pagan traditions.
27. Genesis 6:17–19.
28. Moltmann 1985, 187.
29. Friedrich and Kittel 1985, 1344.
30. Ibid. Genesis: 27– 25; Jeremiah. 3:11.
31. Hallé (2002) has also drawn attention to this lack of plants on Noah's ark.
32. I owe this point entirely to Sally Alsford.
33. For a contemporary angle on the vulnerability of plants to flooding, see Kozlowski 1997 and Jackson and Colmer 2005.
34. Genesis 7:22–23.
35. Some Bible editions translate this as cypress wood.
36. Page 1992, 26; Hillel 2006.
37. Genesis 2:5–9.
38. Spero 1983, 75.
39. Moltmann 1985, 187.
40. Genesis 9:3–4.
41. See Chapters 4, 5 and 6.
42. Genesis 1:29–31.
43. Primavesi 2001, 129.
44. Moltmann 1985, 31.

45. This hierarchy is also expressed in the New Testament. Jesus recognized the beauty of flowering plants, but counsels his apostles that God has much more regard for human life than he has for the lives of plants. "Consider how the lilies grow. They do not labor or spin. Yet I tell you, not even Solomon in all his splendor was dressed like one of these. If that is how God clothes the grass of the field, which is here today, and tomorrow is thrown into the fire, how much more will he clothe you, O you of little faith!" Luke 12:27–28.
46. Primavesi 2001, 129.
47. Amos 2:9.
48. Song of Solomon 2:3.
49. Isaiah 40:6–7.
50. Job 5:25.
51. Psalms 92:8.
52. Job 14:1.
53. Job 29:19: "I open my root towards water. The dew of the night is in my boughs."
54. Jeremiah 17:8.
55. Isaiah 35:1–2.
56. Psalms 96:11–13.
57. Zechariah 11:2.
58. Nielsen 1989, 71.
59. Hillel 2006.
60. Zohary 1982, 28.
61. Jeremiah 2:21: "How then have you turned degenerate and become a wild vine?"
62. Nabhan 1995, 96.
63. Ibid.
64. See Zohary 1982. In contrast, nonagricultural plants are barely mentioned. When they are, as the next section will demonstrate, they are employed almost entirely symbolically as allegories, metaphors, and similes.
65. Hillel 2006.
66. Deuteronomy 8:7–9. My emphasis on the description of plants. The reference to honey is thought to refer to honey from the date palm, *Phoenix dactylifera* L.
67. Ezekiel 36:8.
68. Amos 9:13.
69. Deuteronomy 20:19–20.
70. Luke 13:7–8.
71. Mark 11:14.
72. "I struck you with blight and mildew; your many gardens and your vine-

yards, your fig trees and your olive trees the locust devoured; yet you did not return to me, declares the lord." Similarly, in Exodus 10:15, in order to punish the Pharaoh, God sends down a plague of locusts to eat all the plants of Egypt. "They covered the face of the whole land, so that the land was darkened, and they ate all the plants in the land and all the fruit of the trees that the hail had left. Not a green thing remained, neither tree nor plant of the field, through all the land of Egypt." Amos 4:9.

73. Jonah 4:10–11.
74. Russell 1984, 335.
75. For a discussion of relationality in Augustine's theology, see Jenson 2006.
76. Augustine, *City of God* 12.28; Kenny 1973, 1.
77. Augustine, *City of God* 5.11.
78. This appears to coincide with the idea of Plato in the *Republic* that the soul was comprised of a superior part and an inferior part. *Republic* 602cd.
79. Augustine, *City of God* 5.11.
80. In Book 12, Augustine gives another account of the nature and origin of the human soul "God, then, made man in His own image. For He created for him a soul by virtue of which he might surpass in reason and intelligence all the creatures of the earth, air and sea, which do not have souls of this kind." *City of God* 12.24.
81. Augustine, *City of God* 12.24.
82. In acknowledging that plants are alive and have the faculties of growth and reproduction, Augustine departs from the terminology of Aristotle by not labelling these faculties as facets of soul, but instead deeming them as basic faculties of life.
83. Augustine, *City of God* 11.27.
84. Augustine, *City of God* 1.20.
85. Augustine, *City of God* 8.5.
86. Russell 1984.
87. In the treatment of the living world in *Summa Theologica*, animals were considered to be more "noble" than plants, an idea that can be found echoed in John Ray's *Wisdom of God* several hundred years later. Animals were nobler because they clearly bore greater physical resemblance to human beings, and in spiritual terms, they were more closely related to humans by virtue of the fact that humans and animals were created on the same day. *ST* III, Q.44, A.4. Obj.1.
88. *Summa Theologica (ST)* II–II, Q.179, A.1.
89. *ST* I–II, Q.5, A.1. Obj.1.
90. *ST* Ia. Q.75, A.2.
91. *Summa Contra Gentiles* 2.82.20.
92. *Summa Contra Gentiles* 2.82.9.

93. *Summa Contra Gentiles* 2.82.8.
94. *Summa Theologica* considers animals to be more "noble" than plants.
95. Augustine, *City of God* 1.20 quoted in *ST* II–II, Q.64, A.1. Obj 3.
96. *ST* II–II, Q.64, A.1. Obj 3.
97. Augustine, *City of God* 1.20 quoted in *ST* II–II, Q.64, A.1. Repl Obj 1.
98. *ST* II–II, Q.64, A. 1. Repl Obj 2 "Dumb animals and plants are devoid of the life of reason whereby to set themselves in motion; they are moved, as it were by another, by a kind of natural impulse, a sign of which is that they are naturally enslaved and accommodated to the uses of others."
99. See Introduction.
100. Nash 1991, 77. In addition, as Chapter 2 notes, the biblical voice has wielded particular influence on Western botany. Many of the West's greatest botanists from Nehemiah Grew to Carolus Linnaeus, were devout Christians, and until Darwin published his theory of evolution by natural selection, there was little opposition in science to the creation stories featured in the Bible. Indeed, Linnaeus famously postulated that no new species had come into existence since the Garden of Eden.

Chapter 4. Dealing with Sentience and Violence in Hindu, Jain, and Buddhist Texts

1. *Acaranga Sutra* 1 2.3.4. Quotations from Jain Sutras are from Jacobi 2001.
2. King 1999. Brockington 1996.
3. King 1999, 45.
4. Again, in must be emphasized that I am not attempting to characterize a definitive Hindu position, nor indeed an attitude common to Hindu, I am merely examining plant human dynamics and the points and processes which influence plant-human relationships. My main focus here is to contrast instances of connection and moral inclusion with the Western attitudes found in Chapters 1 to 3.
5. Callicott 1997.
6. As there is no significant difference between the Digambara and the Svetambara position on human-plant relationships, I will treat the Jain tradition as being univocal in this chapter.
7. Callicott 1997, 55–57.
8. Rolston 1987.
9. Such a treatment of plants using Hindu sources is intended to provide a philosophical contrast to, rather than an historical basis for, the Jain and Buddhist traditions.
10. Upadhyaya 1964, 28.
11. Upadhyaya 1964, 18–29.

12. In India, it is said that these deities often protect the trees that they inhabit and need to be consulted about any destructive actions concerning their trees. Where a deity or deities resides in a stand of trees, such areas are usually designated as sacred groves. Grodzins Gold and Gujar 1989.
13. Upadhyaya 1964, 28.
14. Sensarma 1989, 8–9.
15. Sensarma 1989, 40.
16. Lincoln 1986.
17. Lincoln 1980, 3.
18. *Mahābhārata* 12.182.17.
19. *Chāndogya Upaniṣad* 3.14.1.
20. Dwivedi 2000, 4.
21. *Mahābhārata* 12.285. (Ganguli 2000, 323). As the available English translation of Book 12 part 2 does not contain line references, the relevant page numbers are cited instead.
22. *Mundaka Upaniṣad* 2.17–9.
23. Moltmann 1985; Brockington 1992, 1–3.
24. *Bhagavadgītā* 15.7–8; Rao 1970, 380. Although contained within the *Mahābhārata*, the *Bhagavadgītā* is treated separately here. The translation for this quote comes from the Gita Society. Van Buitenen (1981) has also been used in the interpretation of the text.
25. *Mahābhārata* 3.203.33.
26. *Bhagavadgītā* 15.8–9.
27. *Bhagavadgītā* 15.7–8.
28. *Mahābhārata* 12.211.1.
29. *Mahābhārata* 3.199.20.
30. *Mahābhārata* 12.184. (Ganguli 2000, 26).
31. *Vogavasistha* 62.7.
32. The "fivefold" immovable creation refers to plants, the five orders of which are: one, trees; two, shrubs; three, climbing plants; four, creepers; and five, grasses. *Vishnu Purana* 1.5.
33. Fundamental to this idea of rebirth is karma, "the 'doctrine' or 'law' that ties actions to results and creates a determinant link between an individual's status in this life and his or her fate in future lives." The basis of the term is the Sanskrit root word *karman* which literally means "action," and at death, the actions in this life drive the *jiva* into another birth. As the accumulation of karma is difficult to avoid, the process of death and rebirth is an oft-repeated cycle. The process of rebirth is driven by ignorance, and this cycle of *samsara* is only broken by the dissolution of karma, which results in liberation (*moksha*) from the cycle of birth and death. Tull 2004, 309.
34. *Mahābhārata* 12.212. (Ganguli 2000, 95).
35. *Brahadāraṇyaka Upaniṣad* 3.12.13.

36. *Brahadāranyaka Upaniṣad* 6.2.16.
37. *Chāndogya Upaniṣad* 10.10.6.
38. *Manu Smrti* 1.49.
39. *Mahābhārata*. 12.268. (Ganguli 2000, 257). Even sacrifice of these creatures is framed in terms of benefit, for it is stated that this is the only way for them to attain heaven.
40. *Mahābhārata*. 3.198.69.
41. *Mahābhārata*. 3.206.80. My emphasis.
42. *Yoga Sutra of Patanjali*, quoted in Devaraja 1976, 439.
43. Ibid.
44. *Mahābhārata*. 3.199.20–25.
45. Nelson 2000, 142.
46. Ibid. Callicott 1997 also raises similar doubts as to the effectiveness of Vedānta philosophy, characterizing it as world-denying.
47. *Bhagavadgītā* 18.17.
48. As Nelson recognizes, it is stressed in the *Bhagavadgītā* that just as pain and pleasure matter to oneself, so they are matter to all other beings. *Bhagavadgītā* 6.32.
49. Instead of the *Vedas*, Jainism is based upon the teachings of twenty-four spiritual teachers known as the Tirthankaras. The last of the Tirthankaras was Mahavira who lived in Northeast India during the time of the Buddha (sixth century BCE). See Dundas 1992.
50. In the English translation, Mahavira also attributes plants with reason, which is interpreted as an awareness of the environment around them for growth and reproduction. *Acaranga Sutra* 1.1.5.6.
51. *Acaranga Sutra* 1.1.5.6.
52. Ram-Prasad 2007, 45–50.
53. Ram-Prasad 2007, 4.
54. *Acaranga Sutra* 1.8.1.5–6.
55. *Bhagavati Sutra* quoted in Wiley 2002, 41.
56. *Tattvarta Sutra* 2.8. However, this sentient and knowledgeable soul is covered by a blanket of ignorance, as each soul is embedded within a cycle of rebirth that is driven by the karma accrued in previous lifetimes. As in Hinduism, only the dissolution of karma will result in a soul's release from a world of suffering. This liberation (*moksa*) returns the soul to a state of *paramatman* (a divine state often equated with the term *God*). In this sense, the Jain conception of soul is very similar to the Hindu. See Dundas 1992, 110.
57. According to Jainism, there are four destinies for the soul after death. The main ones are heavenly beings, humans, hell beings, animals, and plants. It is also possible for a soul to become embodied in the elements of earth, fire, air, and water—although not all the elements are embodied. As possible

repositories of the soul, all these beings are considered to be alive. Wiley 2002, 39.
58. *Bhagavati Sutra*, cited in Dundas 1992, 106.
59. Cort 2002, 84.
60. Ram-Prasad 2007.
61. Wiley 2002, 42.
62. Ibid.
63. *Uttaradhyayana* 36. 71–106 and *Tattvarta Sutra* 2.12.
64. Note that Jainism classifies the elements as living beings.
65. *Tattvarta Sutra* 2.23.
66. *Tattvarta Sutra* 2.2.4–5. Here, even within this philosophy of connection, there is evident zoocentrism in the use of the criteria for classifying beings.
67. By practicing *ahimsa*, the Jaina cleanses karma from the *jiva*, which leads to the attainment of spiritual liberation.
68. *Tattvarta Sutra* 2.2.5.
69. However, this slight zoocentric appraisal is also very different to that found in the work of Plato and Aristotle. It does not form part of the justificatory framework of a process of backgrounding, that is itself part of the preparation for domination of the natural world. Although it undervalues plant faculties and capabilities, it does not conspire to treat plants as lacking in autonomy and purpose.
70. Ram-Prasad 2007.
71. Ram-Prasad 2007, 40–45.
72. *Acaranga Sutra* 1.2.3.4
73. Accumulated karma binds beings within the *samsaric* world.
74. Ram-Prasad 2007, 4.
75. *Gommatasara Jivakanda*, quoted in Wiley 2002, 57.
76. *Acaranga Sutra* 1.4.2.5
77. It must be remembered that the Jaina definition of life is not identical to the common scientific understanding, for in Jain texts, elements such as water are treated as living and so deserving of *ahimsa* and appropriate respect also. See Chapple 2002.
78. *Acaranga Sutra* 1.1.5.6–7.
79. *Acaranga Sutra* 2.1.8.3. Also, fruits from the genus *Ficus* are prohibited as they contain innumerable tiny insects, while pomegranates are not to be eaten because their consumption necessitates the destruction of many seeds, each with their own *jiva*. *Acaranga Sutra* 2.1.8.6 and Williams 1963, 53.
80. Dundas 1992, 189.
81. Dundas 1992, 191 and Wiley 2002, 52.
82. Callicott 1997.
83. Wiley 2002, 48.
84. *Acaranga Sutra* 1.3.3.1

85. *Sutrakritanga* 2.1.48.
86. The ideal is nonviolence, but even basic acts such as walking and washing result in death for numerous beings.
87. Shilapi 2002, 164.
88. Even plants that form a large part of the human diet, (plants that traditionally in the West we view only as food) must be considered outside of a use-based framework. *Acaranga Sutra* 2.4.2.13–16.
89. *Acaranga Sutra* 2.4.2.12.
90. Ram-Prasad 2007, 9.
91. "The first position holds that Buddhist environmentalism extends naturally from the Buddhist worldview; the second that the Buddhist worldview does not harmonize with an environmental ethic. The third position maintains that one can construct a Buddhist environmental ethic, though not co-terminus with the Buddhist worldview, from Buddhist texts and doctrinal tenets; the fourth, that one should evaluate a viable Buddhist environmental ethic in terms of Buddhist ethics rather than inferred from the Buddhist worldview. The fifth position holds that the most effective Buddhist environmental ethic takes its definition in terms of particular contexts and situations." Swearer 2006, 124–125.
92. See Swearer 2006.
93. Cooper and James 2005, 110. My emphasis.
94. Harris 1991, 104.
95. Cooper and James 2005, 110.
96. The wheel is divided into six inner segments, within which resides a category of sentient beings. These six realms of *samsaric* life are gods, demi gods, human beings, animals, hungry ghosts, and hell beings. The first of the three categories are regarded as "upper" and the latter three as "lower" destinies of birth. Like Jainism, Buddhism rejected the authority of the Brahmins and the pantheon of Hindu gods, but retained the ideas of karma and rebirth. So it too has as a strong focus on the practice of *ahimsa*.
97. *Majjhima Nikāya* 12.35. In subsequent schools of Buddhism, another realm of Gods has been added to make six realms of life.
98. *Majjhima Nikāya* 12.32–34. This is a marked change from the position of the *Upaniṣads*, which include "a category of "sprout-born" beings. *Chāndogya Upaniṣad* 6.3.1, *Aitareya Upaniṣad* 3.1.3. See Findly 2002, 253.
99. In most schools of Buddhism, there is no belief in an unchanging divine soul present in all living beings—a soul that transmigrates at the time of death as in Hinduism and Jainism. It is the dynamic relation of psychic constructs and consciousness that link successive births in the Buddhist understanding. As the second and third link of the twelve links of dependent origination, the karmically shaped mental constructs and consciousness

pass from one body to the next in a continual process of change. These twelve links appear on the outer rim of the Bhavachakra. The description corresponds to a fundamental Buddhist teaching of nonself (*anatman*), which asserts that in the five aggregates no independently existent, immutable self or soul can be found. All phenomena arise in interrelation and in dependence on causes and conditions, and, thus, are subject to inevitable decay and cessation.

100. *Pattimokkhasutta, Pacittiya* 11, quoted in Schmithausen 1991, 5.
101. *Pana* literally means "breathing beings."
102. Schmithausen 1991, 13.
103. Specifically as living beings which had the sense faculty of touch. Schmithausen notes that in popular Buddhism in Burma and Sri Lanka, plants are considered as living beings with one sense faculty. Schmithausen 1991, 13.
104. Schmithausen 1991, 11.
105. Schmithausen 1991, 8.
106. Schmithausen 1991, 20.
107. Schmithausen 1991, 76.
108. Such a stance toward plants is corroborated by another old Buddhist text in which the plants are regarded as being among the different species of animate beings. Schmithausen 1991, 66.
109. Schmithausen 1991, 23.
110. Schmithausen 1991. 16–17.
111. Findly 2002, 256.
112. Ibid.
113. Ibid.
114. Ibid.
115. This is equally true for the monastic and the lay Buddhist. The monastic buildings of earliest Buddhism would have necessitated the clearing of forest, and would have been an important building material for temples and monasteries. Many of the major monasteries that were donated to the Buddhist sangha were located in forest groves—e.g., Veluvana, Jetavana, and the Mahavana monastery in the Sakya capital Kapilavattu.
116. Schmithausen 1991, 73–74.
117. The significance of this for virtue ethics will be discussed later.
118. Schmithausen 1991, 79.
119. Mādhyamaka is a Mahāyāna Buddhist school, which although Indian in origin is now retained chiefly within the body of Tibetan Buddhism. See Snellgrove 1987.
120. Schmithausen 1991, 89.
121. Schmithausen 1991, 91.

122. Schmithausen 1991b, 91–93.
123. See Harvey 1990, 202; Harvey 2000, 69.
124. Dalai Lama 1992, 26.
125. Khenpo Karthar Rinpoche. Online at www.kagyu.org/kagyulineage/buddhism/int/intqa.php.
126. Cooper and James 2005, 131.
127. Lama Thubten Shenphen (May 2002). Online at www.dharmaling.org/content/view/30/32/1/1/lang,en/.
128. Dalai Lama 1992, 26.
129. Dalai Lama 2005, 3. "If scientific analysis were to conclusively demonstrate certain claims of Buddhism to be false, then we must accept the findings of science and abandon those claims."
130. As the next section argues, such a reversal is not without precedence in the Buddhism. Chinese and Japanese scholars of the Mahayana tradition have questioned the notion of plant insentience over a period of several centuries.
131. Cooper and James 2005, 103.
132. Cooper and James 2005, 101.
133. Of course, this point also applies in the case of the layperson.
134. Ram-Prasad 2007, 9.
135. Ibid.
136. LaFleur 1989, 184.
137. The teaching of universalism, that all beings will eventually become enlightened like the Buddha, is a central tenet of Mahayana philosophy.
138. LaFleur 1989, 184.
139. Eliot 1993, 172.
140. LaFleur 1989, 185.
141. Chan Jan quoted in Lafleur 1990, 135.
142. Eliot 1993, 322–23.
143. Callicott 1997, 93–94.
144. Callicott 1997, 96.
145. Parkes 1997, 113.
146. Kasulis 2004, 8.
147. Swanson 1997, 6.
148. Parkes 1997, 114.
149. Hakeda 1972, 255.
150. LaFleur 1989, 187.
151. LaFleur 1989, 189.
152. Ibid.
153. LaFleur 1989, 190.
154. LaFleur 1990, 139.
155. Findly 2002, 261–63.
156. For a discussion of this see Callicott 1997, 95.

157. Chujin translated and quoted in LaFleur 1989, 192.
158. LaFleur 1989, 195–96.
159. Bashō 1966, 33.
160. Ibid.
161. Donald Swearer would probably classify such a position as "eco-apologist."
162. James 2004, 69.
163. Ibid.
164. Schmithausen 1991a, 24.
165. See Callicott 1997.

Chapter 5. Indigenous Animisms, Plant Persons, and Respectful Action

1. Neidjie 1998, 3.
2. Callicott 1997; Grim 2001; Plumwood 2002; and Mathews 2003.
3. Whitt et al. 2003, 4. It is important to note that Indigenous thought is not separable from practice. Cosmology, ritual practice, and subsistence are all commonly interwoven in what have been termed Indigenous "lifeways." Grim 1998 and Grim 2001, 34.
4. Grim 2001, 51.
5. Deutsch 1970, 80.
6. Harvey 2005a.
7. Henare 2001, 201.
8. In a thought provoking article, Brosius (2001) questions the appropriateness of the use of the term *sacred* in the context of Indigenous ecologies, linking it to the "grammar of conquest." In any case, as alluded to in the Introduction, this chapter will refrain from focussing on "sacred" plants as the literature abounds with works on sacred plant species. See Schultes and Hofmann 1992.
9. Although recent analyses of animist cultures have highlighted personhood, plant personhood has been somewhat in the shadow of animal persons. This chapter will complement such studies by focussing exclusively on plant-human relationships.
10. Sanchez 1999, 207–228.
11. Rose 2005, 295.
12. Bradley 2001.
13. Rose 1992.
14. Rose 1992, 57.
15. Rose 1987, 8.
16. Tunbrige 1988, 48.
17. Berndt and Berndt 1989.

18. Rose 1987, 6–7.
19. Bradley 1988, 12.
20. Rose 1996, 29.
21. Whitt et al. 2003, 4.
22. See Rose 1999.
23. Bradley 2001.
24. Neidjie 1998, 4.
25. Detwiler 1992, 238.
26. Detwiler 1992, 238–39.
27. Detwiler 1992, 239.
28. Morrison 2000, 24.
29. Henare 2001, 202.
30. Erenora Puketapu-Hetet, quoted in Whitt et al. 2003, 5.
31. Roberts and Willis 1998, 45.
32. Seton and Bradley 2004, 213.
33. Henare 2001, 202.
34. Ibid.
35. Dreadon 2002, 6.
36. Nelson 1986, 16.
37. Ibid.
38. Again, there is the motif of the initial humanity, pointing toward ideas of personhood that are explored in the next section.
39. Nelson 1986, 17.
40. Mauze 1998, 238. This creation of human beings from plants has strong parallels with a story in the *Poetic Edda* that humans were created from trees. (See Chapter 6).
41. Mauze 1998, 238.
42. The motifs of tree birth and shared blood have strong parallels with the pagan tales in the next chapter.
43. Neidje 1985, 51.
44. Bird-David 1999; Harvey 2005a.
45. Ibid.
46. Harvey 2005a, Preface xi.
47. Harvey 2006.
48. Hallowell 1960, 27.
49. Hallowell 1960, 36. "The interaction of the Ojibwa with certain kinds of plants and animals in everyday life is so structured culturally that individuals act as if they were dealing with "persons" who both understand what is being said to them and have volitional capacities as well."
50. Descola 1992, 115.
51. Harvey 2005a, Preface xvii.

52. See Harvey 2005a, 15 and Guthrie 1993.
53. Harvey 2005a, Preface xi.
54. Harvey 2005a, 16.
55. See Grim (2001) for a criticism of the naïve, romantic concept that native ecological wisdom is genetically transmitted.
56. Harvey 2005a, 18.
57. As in Hindu and Jain texts, as well as ancestral knowledge, the recognition of personhood is based upon that the personal experience that plants, animals, and humans share the capacity to flourish.
58. Detwiler 1992, 238.
59. Detwiler 1992, 239.
60. Mauze 1998, 239.
61. Rose 1999.
62. Bradley 2001, 298; Rose 1999.
63. Rose 1996, 7.
64. Rose 1999, 178.
65. Rose 1996, 23.
66. Stanner 1979.
67. Randall 2003, 86.
68. Mathews 2003, 79.
69. Mathews 2003, 79.
70. Berndt and Berndt 1989, 57–58.
71. Berndt and Berndt 1989, 62.
72. Bakhtin makes clear that whereas monologue is univocal, dialogue requires two or more voices. See Bakhtin 1984.
73. There must be something about coolabah trees. There is a Yanyuwa story of a large coolabah tree, which stood on the shore. The tree was the Dreaming for the Barracuda. One day a large ship came too close to the tree, and the coolabah reached out and grabbed the ship and fell down.
74. Neidjie 1985, 52.
75. Mauze 1998, 239.
76. Bradley 2001, 183.
77. Rose 2005, 297.
78. Roe and Hoogland 1999, 16–17.
79. Neidjie 1985, 51.
80. Kilham et al. 1986, 208.
81. Stevenson nd, 5.
82. Rose 2005, 299.
83. Rose 2005, 299.
84. Hallowell 1960, 43.
85. Nelson 1986, 21.

86. Nelson 1986, 20.
87. Nelson 1986, 22. Perhaps in the terminology of the "new animism" it would be more appropriate to say that humans are more powerful persons, followed by animals and then plants.
88. Detwiler 1992, 240.
89. Detwiler 1992, 242.
90. Ibid.
91. Ibid.
92. See Introduction.
93. Rose 1999.
94. In Hand 1989, 47–50.
95. "In Aboriginal Australia, humans do not have a monopoly on knowledge: "people are quite explicit in saying that other creatures know things that we don't know because they inhabit regions that we do not inhabit." Rose 2005, 297–98.
96. Bakhtin 1981, 1984.
97. Harvey 2005a.
98. In an account related to animal persons, Fienup-Riordan (2001) notes that Yup'ik Eskimo people regard their predatory relationships with geese as showing them appropriate respect. Harvesting geese results in greater numbers returning the next year, whereas avoiding predation insults these animal persons who go elsewhere.
99. Tawhai 2002.
100. Miranda Morris (personal communication on May 11, 2007). Children in particular are taught never to cut the tree as it will suffer and cry out in pain.
101. Mauze 1998, 239.
102. Tawhai 2002, 244.
103. Whether (from a plant's point of view) the harm that is inflicted on plants can be truly propitiated or not is a separate issue.
104. Rose 2005, 296.
105. Rose 1995, 47.
106. Neidjie 1985, 41.
107. Neidjie 1985, 52.
108. Marvell 1892, 90.
109. Rose 2006.
110. Barreiro 1992. Cited in Whitt et al. 2003, 17.
111. Mauze 1998, 241.
112. Ibid.
113. Ibid.
114. Harvey 2005a, 56.

115. Harvey 2005a, 44.
116. Winter 2000.
117. Harvey 2005a, 44.
118. Harvey 2005a, 55.
119. Mauze 1998, 240.
120. Puketapu-Hetet, quoted in Whitt et al. 2003, 14.
121. Neidjie 1998, 4.
122. Harvey 2005a, 106.
123. Mauze 1998, 241.
124. Rival 2001, 60.
125. Ibid.
126. Rose, James and Watson 2003, 3.
127. Harvey 2005a, 165.
128. Rose 1999, 179.
129. Harvey 2005a, 164.
130. Nancy Daiyi, quoted in Rose 2003, 110.
131. Kathy Deveraux, quoted in Rose 2003, 110.
132. Annie Isaac, quoted in Bradley et al. 2006, 1.
133. Bradley et al. 2006, 9.
134. Rose 1992, 82. Species identification unconfirmed.
135. Ibid.
136. Rose 1992, 85.
137. Rose 1999, 182.
138. The Roman Emperor Constantine's conversion to Christianity in the fourth century initiated its establishment as the dominant religion in Europe, a process completed in the fourteenth century. The magnitude of this change is captured by historian Ronald Hutton who describes it as "the greatest alteration in the religious history of Europe." Hutton 1991, 247.

Chapter 6. Pagans, Plants, and Personhood

1. Lönnrot 1999, 171.
2. It has been argued by Ken Dowden that the use of the word *paganism* to describe their belief systems is "a misnomer" (Dowden 1999, 3). The use of the singular paganism can be misleading because it implies that before Christianity, a singular system of belief was found across the whole of Europe. Here we must be clear that pagans did not hold to strict systems of belief, nor were they a united religious force. Pagans had particular ways of understanding the world that were local and diverse. These were not based on belief in the same way as Christianity; they were more akin to ways of

living handed down from the ancestors. Certainly pagan worldviews were distinct and pluralistic. However, with these caveats in place, we can continue to use the word *paganism* for it importantly acknowledges the similarities that these diverse local worldviews possessed.

3. White referred to it as "the greatest psychic revolution in the history of our culture." White 1967, 1205.
4. White 1967; Nash 1989, 87–112; Katz 1994, 56; and Bauckham 2006, 33.
5. See Chapter 3.
6. Hutton 1991, 247.
7. Much has been written on plant sacredness, and so this chapter will focus on analyzing the actual relationships that pagan peoples had with plants.
8. Here a note on nomenclature may be helpful. In this chapter, a distinction is made between *pagan* and *Pagan*. The lower-case adjective brings together an amorphous assembly of which Pagans are a self-identified, contemporary, subgroup. I owe the clarity of this distinction to Graham Harvey.
9. Although a number of the texts that I will use have actually been written by Christians, the thematic material on plants has great similarity to texts demonstrably written by pre-Christians. Although Christians may well have held similar notions about plants, the important point is that these "pagan" themes differ sharply from biblical attitudes to plants.
10. *Natural History*, in Clifton and Harvey 2004, 20.
11. *Pharsalia* 1.4.50–8.
12. *Annals* 14.29.
13. *Germania* 9.
14. Ross 1992, 37.
15. Green 1992, 231.
16. Ibid.
17. Ross 1992, 36 and Green 1992, 213.
18. *Iliad* 16.230–240; Sophocles *Trach.* 1166f.
19. *Natural History* 12.1–3.
20. Ovid, *Metamorphoses*, 106.
21. Munro-Chadwick 1900, 35.
22. Hutton 1991.
23. Hutton 1991 and Russell 1979.
24. In a biological context, this idea has been expressed in Margulis and Sagan 2000.
25. Lincoln 1986.
26. Ibid.
27. Lincoln 1986, 1.
28. Lincoln 1986, 5.
29. Ibid.
30. Lymington 1996, 6.

31. While this medieval text is generally considered not to be definitively pagan, the theme of metamorphosis/transformation is strongly embedded in much older (demonstrably pagan) material, such as the Greek myths. This interpretation leaves the question of how Christians could copy out such pagan subject matter. A possible answer is that this material was recorded in the manner of folklore, quaint, old but ultimately misguided beliefs—outmoded notions that posed little political or spiritual threat to Christianity.
32. See Chapter 4.
33. Lincoln 1986, 5.
34. *Theogony* 2.116–138.
35. *Hymn V to Aphrodite* 2.249–290.
36. *Aeneid* 8.314–315.
37. I owe these points to Thistleton-Dyer 1994, 16.
38. Ross 1992, 38.
39. Ibid.
40. Ibid.
41. Dowden 1999, 70.
42. White 1967.
43. Bird-David 1999, 69–70.
44. Bird-David 1999 and Harvey 2005a.
45. Ibid.
46. Bird-David 1999, 79.
47. Szerszynski 2005, 33. See Descola 1992.
48. Kohák 1997. See Descola 1992.
49. These themes are found both in the older pagan material and the newer texts that have been written by Christians or transcribed from old oral poetry.
50. Mathews 2003, 80.
51. Harvey 2005a and Harvey 2006.
52. *Library of History* 5.23.2.
53. Apollodorus relates the work of Panyasis (fifth century BCE), epic poet that Smyrna or Myrrha is daughter of Theias, King of the Assyrians. (Apollodorus 1998). While Ovid's version of the story's events is the same, he attributes her parentage to Cinyras (*Metamorphoses*). Apollodorus describes Cinyras as King of Cyprus, but Ovid and later authors describe him as King of the Assyrians.
54. Ovid, *Metamorphoses* 10.597–600.
55. Neidjie 1985 and Chapter 5.
56. *Hymn VI to Demeter* 31–50.
57. Ovid, *Metamorphoses* 8.1066–1068.
58. *Aeneid* 3.30–50.
59. See Thomas 1988.

60. *Eclogues* 5.20–21.
61. See Chapter 5.
62. While the Greek myths are clearly pre-Christian, much of the available Old Norse and Anglo Saxon material was written down by Christians. In this context, it is correct to question whether this work that is traditionally recognized as pagan, actually contains pagan content that can be used in this study of human-plant relationships. One potential way of establishing the "paganness" of relevant plant-based material written down by Christians, is to demonstrate the similarity of work that was written by pre-Christian peoples. In this chapter, we can go some way to achieving this by comparing Christian-authored work with the preceding Greek material. Here, in discussing "authentic paganness," I am not referring to or attempting to describe a single pagan culture. I am merely trying to identify common motifs that appear in materials commonly referred to as pagan, in order to contrast these with the established dualisms and hierarchies in biblical materials. Of course, in dealing with these written materials I am speaking in a general fashion and am not suggesting that self-identified Christians all thought (or indeed think) purely about plants in terms of human-dominated hierarchy.
63. Lymington 1996, 57.
64. *Anglo Saxon Rune Poems*.
65. Ibid.
66. Harvey 2005a and Kohák 1993.
67. As Chapter 3 makes clear, such recognition of plant autonomy is neither found in the Bible, nor in Christian theology.
68. *Nine Herbs Charm* 1.3.
69. Ibid.
70. I owe this observation entirely to Robert Wallis.
71. *Nine Herbs Charm* 7.12.
72. Honko 1990, 24. The *Kalevala* is regarded as a *presentation* because in order to render it into a continuous epic narrative, the source material in the *Kalevala* has undergone a substantial editing process. However, as this orally transmitted traditional poetry was gathered from an extremely remote region, which remained largely illiterate up until the twentieth century, the generally animistic themes and ideas are most likely to have their origins in the Karelian pagan worldview. The scholar, Elias Lönnrot who collated them, believed that they "had been preserved and passed down from one generation to another, retaining their original content but changing their outward form." See Kaukonen 1990, 161.
73. Nabhan 1997. Of course, although Christian writers-readers-listeners would also enjoy such stories for entertainment, unless they were embedded in dialogical relationships with other-than-humans, they possibly would perceive these stories differently from animists.

74. Lemminkäinen is one of the heroes of the *Kalevala*.
75. Lönnrot 1999, 171.
76. Buber 1944, 7–8.
77. Lönnrot 1999, 578.
78. Of course, all human life requires the destruction of plant life.
79. Lönnrot 1999, 188.
80. Ibid.
81. Ibid.
82. Online at www.paganfed.org/about-princ.php.
83. Online at www.druidry.org. My emphasis.
84. Damh, OBOD (personal communication of July 10, 2007).
85. Harvey 1997, 170.
86. Damh, OBOD (personal communication of July 10, 2007).
87. Philosopher Erazim Kohák is sceptical of whether talking to trees can yield intelligible communication, yet he recognizes that human communication is a way of acknowledging the other.
88. Kohák 1993.
89. Potia (personal communication of May 31, 2007).
90. Rose 1999, 183.
91. Cadaran (personal communication of April 29, 2007).
92. Letcher 2003.
93. Ibid.
94. www.thedance.com/wicca101/wtblv.htm.
95. Paterson 2004, 359.
96. Paterson 2004, 359.
97. Grimes 2002.
98. Harvey 2005b.
99. www.druidnetwork.org/environment/action/plantingtrees.html.
100. Katz 1997, Mathews 2004, Mabey 2007.
101. www.shadowthreadwitch.wordpress.com/2006/09/28/plants-are-not-medicines.

Chapter 7. Bridging the Gulf

1. Baluška et al. 2005, 31.
2. Plumwood 2006.
3. Plumwood 2006.
4. In his introduction, Hales states that there is "in many respects, a great analogy between plants and animals." Hales 1727, 1.
5. Whippo and Hangarter 2006, 1111.
6. Ibid.

7. Darwin 1880, 572.
8. Darwin 1880, 573.
9. Sachs 1890, 554–55.
10. Tropic movements are directional movements which involve growth facilitated by hormones, whereas nastic movements are often rapid, reversible responses to nondirectional stimuli.
11. Stahlberg 2006a.
12. The discovery of electricity in animals was discovered almost one hundred years before by Luigi Galvani in 1791. See Stahlberg 2006a, 6. An action potential is a change in the electrical potential that travels along a cell's surface with constant speed and magnitude.
13. Darwin 1875, 234–61.
14. Darwin 1875, 173.
15. "It has often been vaguely asserted that plants are distinguished from animals by not having the power of movement. It should rather be said that plants acquire and display this power only when it is of some advantage to them; this being of comparatively rare occurrence, as they are affixed to the ground, and food is brought to them by the air and rain." Darwin 1875a, 206.
16. Darwin 1875, 572.
17. Stahlberg 2006a, 7.
18. Ibid.
19. Tudge 2005, 267.
20. Stahlberg 2006b, 5.
21. Stahlberg 2006a, 7.
22. Ibid.
23. Whippo and Hangarter 2006, 1111.
24. Braam 2005, 373.
25. Baluška, Mancuso, and Volkmann 2006, Preface vi.
26. Izaguirre et al. 2006.
27. Ibid.
28. Trewavas 2002.
29. Trewavas (personal communication on January 11, 2007).
30. Stenhouse 1974.
31. Trewavas 2003.
32. Hort 1938, 201–202.
33. Bradshaw and Hardwick 1989.
34. Phenotypic plasticity is made possible by the modular structure of plants which consists of a leaf, a bud and a below ground root meristem. Restricted to the site of germination, this basic "module" can be put together in a wide range of complex structures with great variations in size, shape, and complexity. Flexible in growth and development, plants thus have the ability to adapt their phenotype in response to signals from a

highly variable and heterogeneous local environment in order to maximize their growth, survival, and reproduction.
35. Trewavas 2003, 13.
36. Trewavas (personal communication on January 11, 2007).
37. Other examples of phenotypic plasticity can be physiological and morphological. Physiologically plastic responses include plasticity in transpiration and rates of photosynthesis. See Bloom et al. 1985. Morphological plasticity is commonly expressed in growth habit and size, morphology and anatomy of vegetative and reproductive structures, in biomass accumulation, growth rates, and sex expression. The genotype determines whether the individual phenotype or character can be plastic; the expression and extent of that plasticity is a function of a plant's perception and assessment of its environment. Sultan 1996; Sultan 2000; and Bradshaw and Hardwick 1989.
38. Forde 2002 and Callaway 2003.
39. Trewavas 2003, 3.
40. Jackson and Caldwell 1996.
41. Hutchings and deKroon 1994.
42. Hutchings and deKroon 1994 and Trewavas 2003.
43. Mahall and Calloway 1991. This is also demonstrated by tree species. Studies of blue oak *Quercus douglasii* Hook. & Arn. in California have shown that genetically identical individuals can produce root systems with architecture that varies greatly. Trees able to access the water table directly develop long tap roots and have very few shallow roots, whereas species that are not able to access the water table have many shallow roots and are in competition for water resources with understorey woodland species. Where possible, the roots will avoid this competition by hitting the water table. Callaway, et al. 2003.
44. Firn 2004.
45. Trewavas 2004.
46. Firn compares the intelligence of plant life to the workings of an automated thermostat. Firn 2004.
47. Struik et al. 2008, 366.
48. Trewavas 2003.
49. Baillaud 1962.
50. Trewavas 2003 and Trewavas 2005.
51. Trewavas 2003, 5. My emphasis.
52. Kelly 1992.
53. Runyon, Mescher, and De Moraes 2006.
54. Plant neurobiology harks back to Darwin who wrote of the "brains" of root caps, which he found to be aware of the environment and to be able to control the movement of other parts. See Brenner et al. 2006.
55. Baluška, Mancuso, and Volkmann 2006, 28.

56. Baluška, Mancuso, and Volkmann 2006.
57. Baluška, Volkmann, and Menzel 2005.
58. Auxin is involved in phototropic and gravitropic responses when transmitted transcellularly. The *extracellular* transmission helps pass rapid information, which controls such vital physiological processes as photosynthesis and respiration, which can be subject to plasticity in response to prevailing environmental conditions. Baluška, Volkmann, and Menzel 2005, 108. In the words of Bateson (1972, 315), this "information is a difference that makes a difference."
59. Baluška, Volkmann, and Menzel 2005.
60. Trewavas 2003, 6.
61. Alpi et al. 2007.
62. Brenner et al. 2007.
63. Barlow 2008.
64. Trewavas 2005.
65. Brenner et al. 2005.
66. The network driven organization of biological life suggests that "there is no single controller." Noble 2006, 113.
67. Noble 2006.
68. Thompson 2007, 60.
69. Capra 1996, 32.
70. Struik et al. 2008.
71. Interestingly, self-recognition is not directly disputed by Struik et al. (2008), simply the fact that this may be labelled as a facet of plant intelligence.
72. Briggs and Walters 1997.
73. Gruntman and Novoplansky 2004.
74. *B. dactyloides* was found to respond differently when faced with competition between self and nonself. In the presence of other roots of the same plant, individuals developed fewer and shorter roots. In the presence of the roots of another individual, plants were shown to act in two different ways. Genetically identical plants that were grown together were found not to compete. Whereas genetically identical plants that had been separated for up to sixty days were found to proliferate their roots when coming into contact with each other. It seemed that the longer that genetically identical plants spent apart, the less they recognized each other as belonging to *self*. On this basis, a simple allogen recognition system was discounted. Gruntman and Novoplansky 2004.
75. Gruntman and Novoplansky 2004, 3865–66.
76. Gruntman and Novoplansky 2004, 3866.
77. Ibid.
78. Ibid.

79. Baluška, Volkman, and Menzel 2005, 110.
80. Gersani et al. 2001.
81. Gruntman and Novoplansky 2004.
82. Bais et al. 2004, 1.
83. Estabrook and Yoder 1998.
84. Peters, Frost and Long 1986.
85. Bais et al. 2004.
86. De Weert et al. 2002.
87. These are known from experimental plants *Arabadopsis thaliana* (L.) Heynh., *Ocimum basilicum* L., *Oxalis tuberosa* Molina, and *Centaurea maculosa* Lam. Bais et al. 2004.
88. Bais et al. 2004.
89. Park et al. 2002.
90. Bais et al. 2004.
91. Callaway 2002 and Bais et al. 2004 .
92. Mahall and Callaway 1992.
93. Quayyum et al. 2000.
94. Evans and Cain 1995.
95. Mahall and Callaway 1992.
96. Falik et al. 2003.
97. Estabrook and Yoder 1998.
98. Dicke and Bruin 2001.
99. Pichersky and Gershenzon 2002.
100. Schiestl et al. 1999.
101. Pichersky and Gershenzon 2002.
102. Berkov, Meurer-Grimes, and Purzycki 2000.
103. Kost and Heil 2006.
104. Dicke and Bruin 2001.
105. Kost and Heil 2006.
106. Ibid.
107. Ibid.
108. Karban et al. 2000; Tscharntke et al. 2001; and Kost and Heil 2006.
109. Peng et al. 2005.
110. Peng et al. 2005 and Kost and Heil 2006.
111. Dolch and Tscharntke 2000.
112. Dudley and File 2007.
113. Struik et al. 2008.
114. Brenner et al. 2007, 286.
115. See Chapter 4.
116. Trewavas (personal communication on January 11, 2007).
117. Ram-Prasad 2007, 9.

Recreating a Place for Flourishing

1. Blake 1972, 151.
2. Blake 1972, 793.
3. Bookchin 1982.
4. Plumwood 1993, 71.
5. Plumwood 1999.
6. Singer 1980, quoted in Plumwood 1999, 200.
7. Warren 2000, 76.
8. Kohák 1993, 384.
9. Kohák 1993, 383.
10. Kohák 1993, 385.
12. My understanding of flourishing here connects with Chris Cuomo, who links flourishing with self-created well-being. Beings that are capable of flourishing are those that have "dynamic charm," a "diffuse 'internal ability' to adapt to or resist change" and for Cuomo, individuals (and their aggregates) flourish when their integrity is maintained, when they are in physical health, and when they are connected in communities that enable mutual opportunities for life and health (Cuomo 1998, 71–74).
13. Plumwood 1993; Warren 2000; and Heyd 2007.
14. Merchant 1982, 9.
15. Plumwood 2002, 26.
16. Plumwood 2002, 27.
17. UNEP 2007 and SCBD, 4.
18. Tilman et al. 2001.
19. Ramancutty and Foley 1999.
20. UNEP 2007, 82 and Laurance et al. 2001.
21. SCBD 2002, 4; Thomas et al. 2004: and Pimm and Raven 2000.
22. IPCC 2001, Section 3.2.4.
23. Neidjie 1985, 52.
24. See Francione 2008.
25. Plumwood 1999, 204.
26. For the importance of ritualization, see Grimes 2002.
27. Although I have sought to frame the differences between Indigenous thought and animal rights theory, his principle actually connects with one of the central tenets of animal welfare. The difference is in the negotiation of what is necessary and acceptable.
28. Baxter and Montgomery 1996, 24.
29. Rose 2005.
30. Bonfante-Fasolo et al. 1981 and Isaac 1992, 318. In order to elicit these associations, it is known that plants release secondary metabolites from their roots, usually a specific combination of metabolites for communication

with particular fungal and microbial species. This communication allows the mutualists to distinguish the roots of their symbiotic partners from those of other plant species and facilitates their mutualistic interaction. Bais et al. 2004.
31. Ritchie 1956, 298.
32. Ritchie 1956, 297.
33. Storch 1993 and Welch et al. 1997.
34. In plants, it is important to note that dialogue is not always concerned with mutual benefit. In many instances, dialogue between plant individuals and species also involves competition for resources.
35. Nabhan 1995 and Harvey 2005a.
36. As the Chapter 6 demonstrates, Western societies already have a significant body of stories that articulate the personhood of plants.
37. Grimes 2002, 157.
38. Rose 1999, 185.
39. Heyd 2007, 32.
40. Kohák 1993, 376.
41. Cuomo 1998, 78.
42. Thompson 2007, 391.
43. In fact, as the very concept of species depends upon the aggregation of individuals, awareness of our effect on individual plants is fundamental. Of course, there are many concepts of species, but the most commonly used—the *biological*, the *taxonomic*, and the *phylogenetic*—all rely on the aggregation of similar individuals.
44. Rose 2006.
45. Chan 2007.
46. WRAP 2007.
47. Ibid.
48. For example, in 2007, the human population of the United Kingdom consumed 11,065,000 m^3 of sawnwood and 12,412,000 tons of paper. To produce this not only requires the felling of millions of trees, each perceptive and autonomous, but it also requires the conversion of natural habitats into plantations. FAO 2007.
49. Ramancutty and Foley 1999.
50. Planta Europa Secratariat 2002, 20.
51. SCBD 2002, 4.
52. Hawkins 2008, 6.
53. Medicinal Plant Specialist Group 2007, 8.
54. McCann 2000.
55. Brooker 2006.
56. Daly et al. 2007.
57. Daly et al. 2007 and Czech et al. 2005.

58. This could also be theorized as a subcategory of both wastage and overconsumption.
59. Steinfeld et al. 2006, Annex 2.
60. De Haan, Steinfeld, and Blackburn 1998, 2.
61. Steinfeld et al. 2006, 12.
62. Ibid.
63. For an example of the effects of soybean cultivation on tropical biodiversity, see Ratter, Ribeiro, and Bridgewater 1997.
64. Steinfeld, et al. 2006, 271.
65. Pahl 2005.
66. Oilseed rape (*Brassica napus* L.), sunflower (*Helianthus annuus* L.), oil palm *Elaeia guineensis* Jacq., and *Elaeis oleifera* (Kunth) Cortés, and soybean (*Glycine max* (L.) Merr).
67. Pin Koh 2007.
68. Pin Koh 2007, 1373.
69. Pin Koh 2007.
70. Searchinger 2008.
71. Ibid.
72. Ibid.
73. On the most local level, leaving space for plants to flourish could begin in our gardens. Along with crops and ornamental plants, dialogue with plants could be instigated by leaving areas of land for the flourishing of local, native species.
74. The balance must be maintained between human use and the needs of plants. This does not preclude some use of protected plant habitats, but it does preclude their total transformation for human ends.
75. Light, 2007, 165.
76. Hettinger 2005; Visvader 1996 cited in Hettinger 2005, 89.
77. Jordan 2003.
78. Grimes 2002, 157. On perfomance and restoration see Jordan 2003.
79. Katz 1997, 105.
80. Katz 1997, 97.
81. Light, 2007, 162–63.
82. Jordan 2003, 96.
83. Jordan 2007, 201.
84. Jordan 2003, 87.
85. Ibid.
86. Rose 2002.
87. Harvey 2005b.
88. Jordan 2003, 78.
89. Jordan 2003, 72.
90. See Plumwood 1993.

91. Rose 2005, 302.
92. Light 2005, 166.
93. Homogenisation is often implicit in Gaian/Deep Ecological philosophies such as Harding 2006, but this is strongly criticized by Plumwood 1993.
94. Mathews 1999.
95. See Berry 2003, 492.

BIBLIOGRAPHY

Alpi, A. et al. 2007. Plant neurobiology, no brain, no gain? *Trends in Plant Science* 12, 135–36.
Anglo Saxon Rune Poem. Online at www.ragweedforge.com/rpae.html.
Apollodorus. 1998. *The Library of Greek Mythology*, translated by R. Hard. New York: Oxford University Press.
Arber, A. 1950. *The Natural Philosophy of Plant Form*. Cambridge, UK: Cambridge University Press.
Aristotle. 1986. *De Anima (On the Soul)*, edited by H. Lawson-Tancred. London: Penguin.
Attfield, R. 2003. Christianity. In *A Companion to Environmental Philosophy*, edited by D. Jamieson, 96–110. Oxford, UK, Blackwell.
Bacon, F. 1670. *Sylva Sylvarum*. London, William Lee.
Baillaud, L. 1962. Mouvements Autonomes des Tiges, Vrilles et Autre Organs. In *Encyclopedia of Plant Physiology, XVII: Physiology of Movements, Part 2*, edited by W. Ruhland, 562–635. Berlin, Springer-Verlag.
Bais, H.P et al. 2004. How Plants Communicate Using the Underground Information Superhighway. *Trends in Plant Science* 19, 26–32.
Bakhtin, M. M. 1981. *The Dialogic Imagination: Four Essays*. Austin, TX: University Texas Press.
Bakhtin, M. M. 1984. *Problems of Dostoevsky's Poetics*. Minneapolis, MN: University of Minnesota Press.
Baluška, F., D. Volkmann, and D. Menzel. 2005. Plant Synapses, Actin-Based Domains for Cell-to-Cell Communication. *Trends in Plant Science* 10: 106–111.
Baluška, F., S. Mancuso, and D. Volkmann, eds. 2006. *Communication in Plants*. Berlin, Springer-Verlag.

Baluška, F. et al. 2006. Neurobiological View of Plants and Their Body Plan. In *Communication in Plants*, edited by F. Baluška et al., 19–36. Berlin: Springer-Verlag.

Barlow, P. W. 2008. Reflections on "plant neurobiology." *Biosystems* 92: 132–47.

Bashō 1966. *The Narrow Road to the Deep North and Other Travel Sketches*, translated by N. Yuasa. Harmondsworth, UK: Penguin.

Bateson, G. 1972. *Steps Toward an Ecology of Mind*. New York: Ballantine Books.

Bauckham, R. 2006. Modern Domination of Nature: Historical Origins and Biblical Critique. In *Environmental Stewardships, Critical Perspectives Past and Present*, edited by R. J. Berry, 32–50. London: T&T Clark.

Bauman, Z. 1993. *Postmodern Ethics*. Oxford, Blackwell.

Baxter, L. A. and B. M. Montgomery. 1996. *Relating: Dialogues and Dialectics*. New York: The Guildford Press.

Barnes, J. 1984. *The Complete Works of Aristotle*. Princeton, NJ: Princeton University Press.

Bekoff, M. 2003. Minding Animals, Minding Earth: Old Brains, New Bottlenecks. *Zygon* 38: 911–41.

Bekoff, M. and J. Goodhall. 2002. *Minding Animals: Awareness, Emotions, and Heart*. New York: Oxford University Press.

Berkov, A, B. Meurer-Grimes, and K. L. Purzycki. 2000. Do Lecythidaceae Specialists Coleoptera Cerambycidae Shun Fetid Tree Species? *Biotropica* 32: 440–51.

Berndt, R. M. and C. H. Berndt. 1989. *The Speaking Land: Myth and Story in Aboriginal Australia*. Ringwood, Australia: Penguin Australia.

Berry, R. J., ed. 2006. *Environmental Stewardships: Critical Perspectives Past and Present*. London: T & T Clark.

Berry, T. 2003. Into the Future. In *This Sacred Earth: Religion, Nature, Environment*, edited by R. S. Gottlieb, 492–96. London: Routledge.

Bhagavadgītā. Online at www.gita-society.com.

Biehl, J., ed. 1997. *The Murray Bookchin Reader*. London: Cassell.

Bird-David, N. 1999. Animism Revisited, Personhood, Environment and Relational Epistemology. *Current Anthropology.* 40: S67–91.

Blake, W. 1972. *Blake, Complete Writings*. Oxford: Oxford University Press.

Bloom, A. J., F. S. Chapin, and H. A. Mooney. 1985. Resource Limitation in Plants—an Economic Analogy. *Annual Review of Ecology and Systematics* 16: 363–92.

Bonfante-Fasolo, P., G. Berta, and V. Gianinazzi-Pearson. 1981. Ultrastructural aspects of endomycorrhizas in the Ericaceae II. Host-endophyte relationships in *Vaccinium myrtillus*. *New Phytologist* 89: 219–244.

Bookchin, M. 1982. *The Ecology of Freedom: The Emergence and Dissolution of Hierarchy.* Palo Alto, CA: Cheshire Books.

Braam, J. 2005. In touch: Plant Responses to Mechanical Stimuli. *New Phytologist* 165: 373–89.
Bradley, J. 1988. *Yanyuwa Country: the Yanyuwa People of Borroloola tell the History of their Land.* Richmond, Australia: Greenhouse Publications.
———. 2001. Landscapes of the Mind, Landscapes of the Spirit: Negotiating a Sentient Landscape. In *Working on Country: Contemporary Indigenous Management of Australia's Lands and Coastal Regions,* edited by R. Baker, J. Davies, and E. Young, 295–307. Oxford: Oxford University Press.
Bradley, J. et al. 2006. *All Kinds of Things from Country: Yanyuwa Ethnobiological Classification.* Aboriginal and Torres Strait Islander Studies Res Rep Series 6. Brisbane, Australia: Aboriginal and Torres Strait Islander Studies Unit, University of Queensland.
Bradshaw, A. D. and K. Hardwick. 1989. Evolution and Stress-Genotypic and Phenotypic Components. *Biological Journal of the Linnaean Society* 37: 137–55.
Brennan, A. 1988. *Thinking About Nature: An Investigation of Nature, Value and Ecology.* Athens, GA: University of Georgia Press.
Brenner, E. D. et al, 2006. Plant Neurobiology: An Integrated View of Plant Signalling. *Trends in Plant Sciences* 11: 413–19.
———. 2007. Response to Alpi *et al.*, Plant neurobiology, the gain is more than the name. *Trends in Plant Science* 12: 285–86.
Briggs, D. and S. M. Walters. 2000. *Plant Variation and Evolution,* 3rd ed. Cambridge, UK: Cambridge University Press.
Brockington, J. L. 1992. *Hinduism and Christianity.* Edinburgh, UK: Edinburgh University Press.
———. 1996. *The Sacred Thread: Hinduism in its Continuity and Diversity.* Edinburgh, UK: Edinburgh University Press.
Brooker, R. 2006. Plant–Plant Interactions and Environmental Change. *New Phytologist* 171: 271–84.
Brosius, J. P. 2001. Local Knowledges, Global Claims: On the Significance of Indigenous Ecologies in Sarwak, Indonesia. In *Indigenous Traditions and Ecology*, edited by J. A. Grim, 125–58. Cambridge, MA: Harvard University Press.
Browning Cole, E. 1992. Theophrastus and Aristotle on Animal Intelligence. In *Theophrastus: His Psychological, Doxographical and Scientific Writings*, edited by W. W. Fortenbaugh and D. Gutas, 44–62. London: Transaction Publishers.
Buber, M. 1944. *I and Thou.* Edinburgh, UK: T. & T. Clark.
van Buitenen, J. A. B. 1975. *The Mahābhārata,* Books 2 and 3. Chicago: University of Chicago Press.

———. 1981. *The Bhagavadgītā in the Mahābhārata*. Chicago: University of Chicago Press.

Callaway, R. M. 2002. The Detection of Neighbors by Plants. *Trends in Ecology and Evolution* 17: 104–105.

Callaway, R. M., S. C. Pennings, and C. L. Richards. 2003. Phenotypic Plasticity and Interactions Among Plants. *Ecology* 84: 1115–128.

Callicott, J. B. 1997. *Earth's Insights: A Multicultural Survey of Ecological Ethics from the Mediterranean Basin to the Australian Outback*. Berkeley, CA: University of California Press.

Callimachus. *Hymn VI to Demeter*. Online at www.theoi.com/Text/Callimachus Hymns2.html#6.

Capra, F. 1996. *The Web of Life*. New York: Anchor Books, New York.

Carone, G. 2003. The Classical Greek Tradition. In *A Companion to Environmental Philosophy*, edited by D. Jamieson. Oxford: Blackwell.

Cashford, J. 2003. *The Homeric Hymns*. London: Penguin.

Chan, K. M. A. 2007. When Agendas Collide, Human Welfare and Biological Conservation. *Conservation Biology* 21: 59–68.

Chapple, C. K. and M. E. Tucker, eds. 2000. *Hinduism and Ecology: The Intersection of Earth, Sky, and Water*. Cambridge, MA: Harvard University Press.

Chapple, C. K., ed. 2002. *Jainism and Ecology: Nonviolence in the Web of Life*. Cambridge, MA: Harvard University Press.

Chaudhur, R. H. N. and D. C. Pal. 1997. Plants in Folk Religion and Mythology. In *Contribution to Indian Ethnobotany*, edited by S. K. Jain, 17–24. Jodphur, India: Scientific Publishers.

Clifton, C. S. and G. Harvey, eds. 2004. *The Paganism Reader*. London: Routledge.

Coleman, F. M. 2006. Book Review. Nature, Technology and the Sacred. *Organization and Environment* 19: 534–35.

Cooper, J. M. 1997. *Plato: Complete Works*. Indianapolis, IN: Hackett.

Cooper, D. E. and S. P. James. 2005. *Buddhism: Virtue and Environment*. Aldershot, UK: Ashgate.

Cort, J. E. 2002. Green Jainism? Notes and Queries Towards a Possible Jain Environmental Ethic. In *Jainism and Ecology: Nonviolence in the Web of Life* edited by C. K. Chapple, 63–94. Cambridge, MA: Harvard University Press.

Cudworth, R. 1678. *The True Intellectual System of the Universe*. London: Printed for Richard Royston.

Cuomo, C. J. 1998. *Feminism and Ecological Communities: An Ethic of Flourishing*. London: Routledge.

Czech, B. et al. 2005. Establishing Indicators For Biodiversity. *Science* 308: 791–92.

Dalai Lama. 1992. *MindScience: An East-West Dialog*. Somerville, MA: Wisdom Publications.
———. 2005. *The Universe in a Single Atom: The Convergence of Science and Spirituality*. New York: Morgan Road Books.
Daly, H. et al. 2007. Are We Consuming Too Much—For What? *Conservation Biology* 21: 1359–62.
Darwin, C. 1875. *Insectivorous Plants*. London: John Murray.
———. 1875a. *The Movements and Habits of Climbing Plants*. London: J. Murray.
———. 1880. *The Power of Movement in Plants*. London: J. Murray.
Deakin, R. 2007. *Wildwood: A Journey Through Trees*. London: Hamish Hamilton.
de Haan, C., H. Steinfeld, and H. Blackburn. 1998. *Livestock and the Environment: Finding a Balance*. Rome, Italy: FAO.
Della Porta, G. 1658. *Natural Magick*. London: Thomas Young and Samuel Speed.
Descola, P. 1992. Societies of Nature and the Nature of Society. In *Conceptualizing Society*, edited by A. Kuper, 107–126. London: Routledge.
Detwiler, F. 1992. All My Relatives: Persons in Oglala Religion. *Religion* 22: 235–46.
Deutsch, E. 1970. Vedānta and Ecology. *Indian Philosophical Annual* 6: 79–88.
———. 1986. A Metaphysical Grounding for Natural Reverence: East-West. *Environmental Ethics* 8: 293–99.
Devaraja, N. K. 1976. What is Living and What is Dead in Traditional Indian Philosophy? *Philosophy East and West* 26: 427–42.
de Weert, S. et al. 2002. Flagella-driven Chemotaxis Towards Exudates Components is an Important Trait for Tomato Root Colonization by *Pseudomonas flourescens*. *Molecular Plant-Microbe Interactions* 15: 1173–80.
Dicke, M. J. and J. Bruin. 2001. Chemical Information Transfer Between Damaged and Undamaged Plants. *Biochemical Systematics and Ecology* 29: 979–1113.
Diels, H. and W. Kranz. 1985. *Die Fragmente der Vorsokratiker*. Zürich, Switzerland: Weidmann.
Diodorus Siculus. 1989. *Library of History*. Volume 5, translated by C. H. Oldfather. London: Loeb.
Dwivedi, O. P. 2000. Dharmic Ecology. In *Hinduism and Ecology: The Intersection of Earth, Sky, and Water*, edited by C. K. Chapple, and M. E. Tucker, 3–22. Cambridge, MA: Harvard University Press.
Dolch, R. and T. Tscharntke. 2000. Defoliation of Alders (*Alnus glutinosa*) Affects Herbivory by Leaf Beetles on Undamaged Neighbours. *Oecologia* 125: 504–511.

Doniger, W. and B. K. Smith. 1991. *Manu-Smriti*. London: Penguin.
Dowden, K. 1999. *European Paganism: The Realities of Cult from Antiquity to the Middle Ages*. London: Routledge.
Dreadon, E. 2002. He Taonga Tuku Iho, Hei Ara: A Gift Handed Down as a Pathway. In *Readings In Indigenous Religions*, edited by G. Harvey, 250–58. London: Continuum.
Dudley, S. A. and A. File. 2007. Kin Recognition in an Annual Plant. *Biology Letters* 3: 435–38.
Dundas, P. 1992. *The Jains*. London: Routledge.
Dyson, R. W. 1998. *Augustine: The City of God against the Pagans*. Cambridge, UK: Cambridge University Press.
Egerton, F. N. 2001. A History of the Ecological Sciences: Early Greek Origins. *Bulletin of the Ecological Society of America*: 93–97, January 2001.
Eliot, C. 1993. *Japanese Buddhism*. Richmond, UK: Curzon Press.
Emel, J., C. Wilbert, and J. Wolch. 2002. Animal Geographies. *Society and Animals* 10: 407–412.
Empedocles. 1981. *The Extant Fragments*, edited by M. R. Wright. New Haven, CT: Yale University Press.
Estabrook, E. M. and J. I. Yoder. 1998. Plant-Plant Communications, Rhizosphere Signaling Between Parasitic Angiosperms and their Hosts. *Plant Physiology* 116: 1–7.
Evans, J. P. and M. L. Cain. 1995. A Spatially Explicit Test of Foraging Behaviour in a Clonal Plant. *Ecology* 76: 1147–55.
Falik, O. et al. 2003. Self/Non-Self Discrimination in Roots. *Journal of Ecology* 91: 525–31.
FAO. 2002. *The State of Food Insecurity in the World*. Rome, Italy: FAO.
———. 2007. *State of the World's Forests, Annex Country Data*. Online at ftp,//ftp.fao.org/docrep/fao/009/a0773e/a0773e10.pdf.
Fathers of the English Dominican Province. 1911–1925. *The Summa Theologica of St. Thomas Aquinas*. London: R & T Washbourne.
Fienup-Riordan, A. 2001. A Guest on the Table: Ecology from the Yup'ik Eskimo Point of View. In *Indigenous Traditions and Ecology*, edited by J. A. Grim, 541–58. Cambridge, MA: Harvard University Press.
Findly, E. B., 2002. Borderline Beings: Plant Possibilities in Early Buddhism. *Journal of the American Oriental Society* 122: 252–63.
Firn, R. 2004. Plant Intelligence: An Alternative Point of View. *Annals of Botany* 93: 345–51.
Forde, B. G. 2002. Local and Long Range Signalling Pathways Regulating Plant Responses to Nitrate. *Annual Review of Plant Biology* 53: 203–224.
Francione, G. L. 2008. *Animals as Persons: Essays on the Abolition of Animal Exploitation*. New York: Columbia University Press.

Friedrich, G. and G. Kittel. 1985. *Theological Dictionary of the New Testament*, translated by Geoffrey W. Bromily. Grand Rapids, MI: Eerdmans Exeter Paternoster.

Garber, D. 2002. Descartes, Mechanics and the Mechanical Philosophy. In *Renaissance and Early Modern Philosophy*, edited by P. A. French and H. K. Wettstein, 185–204. Oxford: Blackwell Publishing.

Gersani, M. et al. 2001. Tragedy of the Commons as a Result of Root Competition. *Journal of Ecology* 89: 660–69.

Green M. J. 1992. *Dictionary of Celtic Mythology and Legend*. London, Thomas and Hudson.

Grim, J. A. 1998. Indigenous Traditions and Ecology. *Earth Ethics* 10, no. 1 (Fall 1998). Online at http://fore.research.yale.edu/religion/indigenous.

Grim, J. A., ed. 2001. *Indigenous Traditions and Ecology*. Cambridge, MA: Harvard University Press.

Grodzins Gold, A. and B. R. Gujar. 1989. Of Gods, Trees and Boundaries: Divine Conservation in Rajasthan. *Asian Folklore Studies* 48: 211–29.

Gruntman, M., and A. Novoplansky. 2004. Physiologically Mediated Self/Nonself Discrimination in Roots. *PNAS* 101: 3863–67.

Guthrie, S. 1993. *Faces in the Clouds: A New Theory of Religion*. New York: Oxford University Press.

Habel, N. 2000. Introducing Earth Bible. In *The Earth Bible Volume 1: Readings from the Perspective of Earth*, edited by N. Habel. Sheffield, UK: Academic Press.

Haddas, M., ed. 1982. *The Complete Plays of Sophocles*. London: Bantam.

Hakeda, Y. S. 1972. *Kūkai: Major Works*. New York: Columbia University Press.

Haldane, E. S. and G. R. T. Ross, eds. 1955. *The Philosophical Works of Descartes*. New York: Dover Publications.

Hales, S. 1727. *Vegetable Staticks*. London: W. and J. Innys, and T. Woodward.

Hallé, F. 2002. *In Praise of Plants*. Portland, OR: Timber Press.

Hallowell, A. I. 1960. Ojibwa Ontology, Behavior and World View. In *Culture in History: Essays in Honor of Paul Radin*, edited by Stanley Diamond, 19–52. New York: Columbia University Press.

Hand, S., ed. 1989. *The Levinas Reader*. Oxford: Blackwell.

Harding, S. 2006. *Animate Earth: Science, Intuition and Gaia*. Totnes, UK: Green Books.

Harris, I. 1991. How Environmentalist Is Buddhism? *Religion* 21: 101–114.

Harvey, D. 1998. *Justice, Nature and the Geography of Difference*. Oxford: Blackwell.

Harvey, G. 1997. *Listening People, Speaking Earth: Contemporary Paganism*. London: Hurst & Company.

———, ed. 2000. *Indigenous Religions: A Companion*. London: Cassell Academic.

———. 2002. *Readings in Indigenous Religions*. London: Cassell Academic.
———. 2005a. *Animism: Respecting the Living World*. London: Hurst & Company.
———. 2005b. An Animist Manifesto. Online at http://www.bioregionalanimism.com/2007/09/animist-manifesto.html.
———. 2006. Animals, Animists, and Academics. *Zygon* 41: 9–20.
Harvey, P. 1990. *An Introduction to Buddhism Teachings, History and Practices*. Cambridge, UK: Cambridge University Press.
———. 2000. *An Introduction to Buddhist Ethics: Foundations, Values, and Issues*. Cambridge, UK: Cambridge University Press.
Hawkins, B. 2008. *Plants for Life: Medicinal Plant Conservation and Botanic Gardens*. Richmond, UK: Botanic Gardens Conservation International.
Healy, J. F. 1999. *Pliny the Elder on Science and Technology*. Oxford: Oxford University Press.
Henare, M. 2001. Tapu, Mana, Mauri, Hau, Wairua: A Maori Philosophy of Vitalism and Cosmos. In *Indigenous Traditions and Ecology*, edited by J. A. Grim, 197–221. Cambridge, MA: Harvard University Press.
Hesiod. *Hymn V to Aphrodite*. Online at www.omacl.org/Hesiod/hymns.html.
———. *Theogony*. Online at www.omacl.org/Hesiod/theogony.html.
Hettinger, N. 2005. Respecting Nature's Autonomy in Relationship with Humanity In *Recognizing the Autonomy of Nature*, edited by T. Heyd, 86–98. New York: Columbia University Press.
Heyd, T. 2007. *Encountering Nature: Toward an Environmental Culture*. Aldershot, UK: Ashgate.
———, ed. 2005. *Recognizing the Autonomy of Nature*. New York: Columbia University Press.
Hillel, D. 2006. *The Natural History of the Bible: An Environmental Exploration of the Hebrew Scriptures*. New York: Columbia University Press.
Homer. 2003a. *The Iliad*, edited by P. Jones. London: Penguin Classics.
———. 2003b. *The Odyssey*, edited by E. V. Rieu. London: Penguin Classics.
Honko, L. 1990. The Kalevala and the World's Epics: An Introduction. In *Religion, Myth and Folklore in the World's Epics: The Kalevala and its Predecessors*, edited by L. Honko, 1–26. Berlin: Mouton De Gryter.
———, ed. 1990. *Religion, Myth and Folklore in the World's Epics: The Kalevala and its Predecessors*. Berlin: Mouton De Gryter.
Hort, A., ed. 1938. *The Critica Botanica of Linnaeus*. London: The Ray Society.
Hutchings, M. J. and H. de Kroon. 1994. Foraging in Plants: the Role of Morphological Plasticity in Resource Acquisition. *Advances in Ecological Research* 25: 159–238.
Hutton, R. 1991. *The Pagan Religions of the Ancient British Isles: Their Nature and Legacy*. Oxford: Blackwell.

Ingold, T. 1994. *What is an Animal?* London: Routledge.
Intergovernmental Panel on Climate Change. 2001. *Climate Change 2001: Synthesis Report.* Cambridge, UK: Cambridge University Press.
Isaac, S. 1992. *Fungal-Plant Interactions.* Berlin: Springer-Verlag.
Izaguirre, M. M. et al. 2006. Remote Sensing of Future Competitors: Impacts on Plant Defences. *PNAS* 103: 7170–74.
Jackson, R. B. and M. M. Caldwell. 1996. Integrating Resource Heterogeneity and Plant Plasticity: Modelling Nitrate and Phosphate Uptake in a Patchy Soil Environment. *Journal of Ecology* 84: 891–903.
Jackson, M.B. and T. D. Colmer. 2005. Response and Adaptation by Plants to Flooding Stress. *Annals of Botany* 96: 501–505.
Jacobi, H. 2001. *Jaina Sutras.* London: Routledge-Curzon.
James, S. P. 2004. *Zen Buddhism and Environmental Ethics.* Aldershot, UK: Ashgate.
Jenson, M. 2006. *The Gravity of Sin: Augustine, Luther and Barthan on Homo Incurvatus in Se.* London: T & T Clark.
Jordan, W. 2003. *The Sunflower Forest: Ecological Restoration and the New Communion with Nature.* Berkeley, CA: University of California Press.
———. 2005. Autonomy, Restoration and the Law of Nature. In *Recognizing the Autonomy of Nature*, edited by T. Heyd, 189–205. New York: Columbia University Press.
Karban, R. et al. 2000. Communication Between Plants: Induced Resistance in Wild Tobacco Plant Following Clipping of Neighboring Sagebrush. *Oecologia* 125: 66–71.
Kasulis, T. P. 2004. *Shinto, the Way Home.* Honolulu, HI: University of Hawai'i Press.
Katz, E. 1994. Judaism and the Ecological Crisis. In *Worldviews and Ecology: Religion, Philosophy and the Environment*, edited by M. E. Tucker and J. A. Grim, 55–70. Maryknoll, NY: Orbis.
———. 1997. *Nature as Subject: Human Obligation and Natural Community.* Lanham, MD: Rowman and Littlefield.
Kaukonen, V. 1990. The Kalevala as Epic. In *Religion, Myth and Folklore in the World's Epics: The Kalevala and its Predecessors*, edited by L. Honko, 157–180. Berlin: Mouton De Gryter.
Kelly, C. K. 1992. Resource Choice in *Cuscuta europaea. PNAS* 89: 12194–197.
Kenny, A. 1973. *The Anatomy of the Soul.* Oxford: Blackwell.
Kilham, C. et al. 1986. *Dictionary and Source Book of the Wik-Mungkan Language.* Darwin, Australia: Summer Institute of Linguistics, Australian Aborigines Branch.
King, R. 1999. *Indian Philosophy: an Introduction to Hindu and Buddhist Thought.* Edinburgh, UK: Edinburgh University Press.

Kohák, E. 1993. Speaking to Trees. *Critical Review* 6: 371–88.

———. 1997. Varieties of Ecological Experience. *Environmental Ethics* 19: 153–71.

Kost, C. and M. Heil. 2006. Herbivore-Induced Plant Volatiles Induce an Indirect Defence in Neighbouring Plants. *Journal of Ecolology* 94: 619–28.

Kozlowski, T. T. 1990. Sattva, Enlightenment for Plants and Trees. In *Dharma Gaia: A Harvest of Essays in Buddhism and Ecology*, edited by A. H. Badiner. 136–44. Berkeley, CA: Parallax Press.

———. 1997. Responses of Woody Plants to Flooding and Salinity. In *Tree Physiology Monograph No. 1*. Online at www.heronpublishing.com/tp/monograph/kozlowski.pdf.

LaFleur, W. R. 1989. Saigyō and the Buddhist Value of Nature. In *Nature in Asian Traditions of Thought: Essays in Environmental Philosophy*, edited by J. B. Callicott and R. T. Ames, 183–209. Albany, NY: SUNY Press.

Laurance, W. F. et al. 2001. The Future of the Brazilian Amazon. *Science* 291: 438–39.

Leclerc, G. L. 1753. History of Man and the Quadrapeds. In P. Hyland, O. Gomez and F. Greensides, eds. 2003. *The Enlightenment: A Sourcebook and Reader*, 107–110. London: Routledge.

Leiss, W. 1972. *The Domination of Nature*. New York: George Braziller.

Letcher, A. 2003. Gaia told me to do it: Resistance and the Idea of Nature Within Contemporary British Eco-Paganism. *Ecotheology* 8: 61–84.

Light, A. 2005. Restoration, Autonomy and Domination. In *Recognizing the Autonomy of Nature*, edited by T. Heyd, 154–169. New York: Columbia University Press.

Lincoln, B. 1986. *Myth, Cosmos, and Society: Indo-European Themes of Creation and Destruction.* Cambridge, MA: Harvard University Press.

Linnaeus, C. 2003. *Systema Naturae*. In *The Enlightenment: A Sourcebook and Reader*, edited by P. Hyland, O. Gomez, and F. Greensides, 102–106. London: Routledge, London. (Originally published in 1735).

Locke, John. 1690. *Essay Concerning Human Understanding*. London: Printed by Elizabeth Holt for Thomas Bassett.

———. 1980. *Second Treatise of Government*, edited by C. B. Macpherson. London: Hackett.

Lönnrot, E. 1999. *The Kalevala: An Epic Poem After Oral Tradition*. New York: Oxford University Press.

Lovejoy, A. O. 1936. *The Great Chain of Being: A Study of the History of an Idea.* New York: Harper.

Low, N., ed. 1999. *Global Ethics and Environment*. London: Routledge.

Lucan. 1956. *Pharsalia: Dramatic Episodes of the Civil Wars*, edited by R. Graves. London: Harmondsworth.

Lymington, C. 1996. *The Poetic Edda*. New York: Oxford University Press.

Mabey, R. 2007. *Beechcombings: The Narratives of Trees*. London: Chatto and Windus.
Mahall, B. E. and R. M. Callaway. 1991. Root Communication Among Desert Shrubs. *PNAS* 88: 874–76.
Margulis, L. and D. Sagan. 2000. *What is Life?* Berkeley, CA: University of California Press.
Marvell, A. 1892. *The Poems of Andrew Marvell*, edited by G. A, Aitken. London: Lawrence & Bullen.
Mathews, F. 1991. *The Ecological Self*. London: Routledge.
———. 1994. *Cultural Relativism and Environmental Ethics*. Originally published in IUCN Ethics Working Group No. 5. Online at www.freyamathews.com.
———. 1999. Letting the World Grow Old: An Ethos of Countermodernity. *Worldviews, Environment, Culture, Religion* 3: 119–37.
———. 2003. *For Love of Matter: A Contemporary Panpsychism*. Albany, NY: SUNY Press.
———. 2004. Letting the World Do the Doing. *Australian Humanities Review* 33. Online at www.lib.latrobe.edu.au/AHR/archive/Issue-August-2004/matthews.html.
Maturana, H. and F. Varela. 1980. *Autopoiesis and Cognition: The Realization of the Living*. Dordrecht, The Netherlands: Kluwer.
Mauzé, M. 1998. Northwest Coast Trees: From Metaphors for Culture to Symbols in Culture. In *The Social Life of Trees*, edited by L. Rival, 233–51. Oxford: Berg.
Mayr, E. 1969. Footnotes in the Philosophy of Biology. *Philosophy of Science* 36: 197–202.
McCann, K. S. 2000. The Diversity-Stability Debate. *Nature* 405: 228–33.
Medicinal Plant Specialist Group. 2007. International Standard for Sustainable Wild Collection of Medicinal and Aromatic Plants (ISSC-MAP). Version 1.0. Bonn, Gland, Frankfurt, and Cambridge. Bundesamt für Natur-schutz (BfN), MPSG/SSC/IUCN, WWF Germany, and TRAFFIC. Online at www.floraweb.de/map-pr.
Merchant, C. 1982. *The Death of Nature*. London: Wildwood House.
Mitra, V. 1976–1978. *The Yogavasistha-Maha-Ramayana of Valmiki*. Varanasi, India: Bharariya Publishing House.
Moltmann, J. 1985. *God in Creation: An Ecological Doctrine of Creation*. London: SCM.
Morrison, K. M. 2000. The Cosmos as Intersubjective: Native American Other than Human Persons. In *Indigenous Religions: A Companion*, edited by G. Harvey, 23–36. London: Cassell Academic.
Morton, A. G. 1982. *History of Botanical Science: An Account of the Development of Botany from Ancient Times to the Present Day*. London: Academic Press.

Mourelatos, A. P. D. 1993. *Pre-Socratics: A Collection of Critical Essays*. Princeton, NJ: Princeton University Press.
Munro-Chadwick, H. 1900. The Oak and The Thunder God. *The Journal of the Anthropological Institute of Great Britain and Ireland* 30: 22–44.
Nabhan, G. P. 1997. *Cultures of Habitat: On Nature, Culture and Story*. Washington, DC: Counterpoint.
Nash, J. 1991. *Loving Nature: Ecological Integrity and Christian Responsibility*. Nashville, TN: Abingdon Press.
Nash, R. 1989. *The Rights of Nature: A History of Environmental Ethics*. Madison, WI: University of Wisconsin Press.
Nee, S. 2004. More Than Meets the Eye. *Nature* 429: 804–805.
Neidjie, B. 1985. *Kakadu Man*. New South Wales, Australia: Mybrood.
———. 1998. Story About Feeling. Broome, Australia: Magabala Books.
Nelson, L. 2000. Reading the *Bhagavadgītā* from an Ecological Perspective. In *Hinduism and Ecology: The Intersection of Earth, Sky, and Water*, edited by C. K. Chapple and M. E. Tucker, 127–64. Cambridge, MA: Harvard University Press.
Nelson, R. K. 1986. *Make Prayers to Raven: A Koyukon View of the Northern Forest*. Chicago: University of Chicago Press.
Niebuhr, H. R. 2006. *The Meaning of Revelation*. Lousville, KY: Westminster John Knox Press.
Nielsen, K. 1989. *There is Hope for a Tree: The Tree as Metaphor in Isaiah*. Sheffield, UK: JSOT Press.
Nine Herbs Charm. Online at www.heorot.dk/woden-9herbs.html.
Noble, D. 2006. *The Music of Life: Biology Beyond the Genome*. New York: Oxford University Press.
Ovid. 2005. *Metamorphoses*, translated by C. Martin. New York: Norton.
Page, R. 1992. The Bible and the Natural World. In *Christianity and Ecology: Seeking the Well-Being of Earth and Humans*, edited by D. T. Hessel and R. Radford Ruether, 20–34. Cambridge, MA: Harvard University Press.
Pahl, G. 2005. *Biodiesel: Growing a New Energy Economy*. White River Junction, VT: Chelsea Green Publishing.
Park, S. W. et al. Isolation and Characterization of a Novel Ribosome-Inactivating Protein from Root Cultures of Pokeweed and Its Mechanism of Secretion from Roots. *Plant Physiology* 130: 164–78.
Parkes, G. 1997. Voices of Mountains, Trees and Rivers: Kūkai, Dōgen and a Deeper Ecology. In *Buddhism and Ecology: The Interconnection of Dharma and Deeds*, edited by M. E. Tucker and D. R. Williams, 111–30. Cambridge, MA: Harvard University Press.
Passmore, J. 1974. *Man's Responsibility for Nature, Ecological Problems and Western Traditions*. London: Duckworth.

Paterson, B. 2004. Finding Your Way in the Woods: The Art of Conversation With the Genius Loci. In *The Paganism Reader*, edited by C. Clifton and G. Harvey, 354–64. London: Routledge.

Peng, J. Y. et al. 2005. Preliminary Studies on Differential Defence Responses Induced During Plant Communication. *Cell Research* 15: 187–92.

Peters, N. K., J. W. Frost, and S. R. Long. 1986. A Plant Flavone: Luteolin, Induces Expression of *Rhizobium meliloti* Nodulation Genes. *Science* 233: 977–80.

Pichersky, E. and J. Gershenzon 2002. The Formation and Function of Plant Volatiles: Perfumes for Pollinator Attraction and Defense. *Current Opinion in Plant Biology* 5: 237–43.

Pimm, S. L. and P. Raven. 2000. Extinction by Numbers. *Nature* 403: 843–45.

Pin Koh, L. 2007. Potential Habitat and Biodiversity Losses from Intensified Biodiesel Feedstock Production. *Conservation Biology* 21: 1373–75.

Planta Europa Secratariat 2002. *European Plant Conservation Strategy*. London: Plantlife International.

Pliny. 1938–1963. *Natural History*, edited by H. Rackham. Ten volumes. London: Heinemann.

Plumwood, V. 1993. *Feminism and the Mastery of Nature*. London: Routledge.

———. 1999. Ecological Ethics from Rights to Recognition, Multiple Spheres of Justice for Humans, Animals and Nature. In *Global Ethics and Environment*, edited by N. Low, 188–212. London: Routledge.

———. 2002. *Environmental Culture: The Ecological Crisis of Reason*. London: Routledge.

———. 2006. Nature in the Active Voice. Unpublished Manuscript.

Primavesi, A. 2001. Ecology and Christian Hierarchy. In *Women as Sacred Custodians of the Earth? Women Spirituality and the Environment*, edited by A. Low and S. Tremayne, 121–41. London: Berghahn Books.

Priscian. 1997. *On Theophrastus on Sense-Perception*, edited by P. Huby. London: Duckworth.

Quayyum, H. A. et al. 2000. Growth Inhibitory Effects of Nutgrass (*Cyperus rotundus*) on Rice (*Oryza sativa*) Seedlings. *Journal of Chemical Ecology* 26: 2221–31.

Ramancutty, N. and J. A. Foley. 1999. Estimating Historical Changes in Global Land Cover. *Global Biogechemical Cycles* 13: 997–1027.

Ram-Prasad, C. 2007. *Indian Philosophy and the Consequences of Knowledge: Themes in Ethics, Metaphysics and Soteriology*. Aldershot, UK: Ashgate.

Randall, B. 2003. *Songman: The Story of an Aboriginal Elder*. Sydney, Australia: ABC Books.

Rao, K. L. S. 1970. On Truth: A Hindu Perspective. *Philosophy East and West* 20: 377–82.

Ratter, J. A., J. F. Ribeiro, and S. Bridgewater. 1997. The Brazilian Cerrado Vegetation and Threats to its Biodiversity. *Annals of Botany* 80: 223–30.

Ray, J. 1798. *The Wisdom of God Manifested in the Works of Creation.* Glasgow, UK: Mundell & Son, and J. Mundell.

Regan, T. 1983. *The Case for Animal Rights.* Berkeley, CA: University of California Press.

Ritchie, J. C. 1956. Biological Flora of the British Isles, *Vaccinium myrtillus* L. *Journal of Ecology* 44: 291–99.

Rival, L. 2001. Seed and Clone: The Symbolic and Social Significance of Bitter Manioc Cultivation. In *Beyond the Visible and the Material: The Amerindianization of Society in the Work of Peter Rivière*, edited by L. Rival and N. Whitehead, 57–80. New York: Oxford University Press.

———, ed. 1998. *The Social Life of Trees.* Oxford: Berg.

Rival, L. and N. Whitehead, eds. 2001. *Beyond the Visible and the Material: The Amerindianization of Society in the Work of Peter Rivière.* New York: Oxford University Press.

Roberts, R. M. and P. R. Willis. 1998. Understanding Maori Epistemology: A Scientific Perspective. In *Tribal Epistemologies: Essays in the Philosophy of Anthropology*, edited by H. Wautischer, 43–78. Aldershot, UK: Ashgate.

Roe, P. and F. Hoogland. 1999. Black and White: A Trail to Understanding. In *Listen to the People, Listen to the Land*, edited by J. Sinatra and P. Murphy 11–30. Victoria, Australia: Melbourne University Press.

Roebuck, V. J. 2003. *The Upanisads.* London: Penguin.

Rolston, H. 1987. Can the East Help the West to Value Nature? *Philosophy of East and West* 37: 172–90.

Rose, B. 1995. *Land Management Issues: Attitudes and Perceptions Amongst Aboriginal People of Central Australia.* Alice Springs: Australia: Central Land Council.

Rose, D. 1987. A Social and Ecological Analysis of Dreaming Trees. Unpublished Manuscript.

———. 1992. *Dingo Makes Us Human: Life and Land in an Aboriginal Australian Culture.* Cambridge, UK: Cambridge University Press.

———. 1996. *Nourishing Terrains: Australian Aboriginal Views of Landscape and Wilderness.* Canberra, Australia: Australian Heritage Commission.

———. 1999. Indigenous Ecologies and an Ethic of Connection. In *Global Ethics and Environment*, edited by N. Low, 175–87. London: Routledge.

———. 2002. *Country of the Heart: An Indigenous Australian Homeland.* Canberra, Australia: Aboriginal Studies Press.

———. 2005. An Indigenous Philosophical Ecology: Situating the Human. *The Australian Journal of Anthropology* 16: 294–305.

———. 2006. What if the Angel of History Were a Dog? *Cultural Studies Review* 12: 67–78.

Rose, D., D. James, and C. Watson. 2003. *Indigenous Kinship with the Natural World in New South Wales.* New South Wales, Australia: NSW National Parks and Wildlife Service.

Rose, D. and L. Robin. 2004. The Ecological Humanities in Action: An Invitation. *Australian Humanities Review.* Online at, www.lib.latrobe.edu.au/AHR/archive/Issue-April-2004/rose.html.

Ross, A. 1992. *Pagan Celtic Britain: Studies in Iconography and Tradition.* London: Constable.

Runyon, J. B, M. C. Mescher and C. M. De Moraes. 2006. Volatile Chemical Cues Guide Host Location and Host Selection by Parasitic Plants. *Science* 313: 1964–67.

Russell, B. 1984. *History of Western Philosophy and Its Connection With Political and Social Circumstances from the Earliest Times to the Present Day.* London: Unwin Hyman.

Russell, C. 1979. The Tree as a Kinship Symbol. *Folklore* 2: 217–33.

Sachs, J. 1890. *History of Botany, 1530–1860.* London: Russell and Russell.

Sanchez, C. L. 1999. Animal, Vegetable, Mineral: The Sacred Connection. In *Ecofeminism and the Sacred*, edited by C. J. Adams, 207–28. New York: Continuum.

Saint Thomas Aquinas. 1955–1957. *Summa Contra Gentiles*, edited by C. J. O'Neil. Notre Dame, IN: University of Notre Dame Press.

Sarton, G. 1952. *A History of Science: Ancient Science Through the Golden Age of Greece.* Cambridge, MA: Harvard University Press.

Schiestl, F. P. et al., 1999. Orchid Pollination by Sexual Swindle. *Nature* 399: 421–22.

Schmithausen, L. 1991a. *Buddhism and Nature.* Tokyo, Japan: International Institute for Buddhist Studies.

———. 1991b. *The Problem of the Sentience of Plants in Earliest Buddhism.* Tokyo, Japan: International Institute for Buddhist Studies.

Searchinger, T. 2008. Use of U.S. Croplands For Biofuels Increases Greenhouse Gases Through Emissions from Land Use Change. Online at www.scienceexpress.org.

Secretariat of the Convention on Biological Diversit.y 2002. *Global Strategy For Plant Conservation.* Montreal, Canada: Secretariat of the Convention on Biological Diversity

Sensarma, P. 1989. *Plants in the Indian Puranas.* Calcutta, India: Naya Prokash.

Seton, K. A. and J. J. Bradley. 2004. "When You Have No Law You Are Nothing": Cane Toads, Social Consequences and Management Issues. *The Asia Pacific Journal of Anthropology* 5: 205–25.

Shilapi, S. 2002. The Environmental and Ecological Teachings of Tirthankara Mahavira. In *Jainism and Ecology: Nonviolence in the Web of Life*,

edited by C. K. Chapple, 159–67. Cambridge, MA: Harvard University Press.
Singer, P. 1980. Animals and the Value of Life. In *Matters of Life and Death*, edited by T. Regan, 218–59. New York: Random House.
———. 1995. *Animal Liberation*. London: Pimlico.
Snellgrove, D. L. 1987. *Indo-Tibetan Buddhism Indian Buddhists and their Tibetan Successors*. London: Serindia.
Soper, K. 1995. *What is Nature?* Oxford: Blackwell.
Soulé, M. and G. Lease. 1995. *Reinventing Nature? Responses to Postmodern Deconstruction*. Washington, DC: Island Press.
Spero, S. 1983. *Morality, Halakha and the Jewish Tradition*. New York: KTAV Publishing.
Stahlberg, R. 2006b. Historical Introduction to Plant Electrophysiology. *Plant Electrophysiology* 1: 3–14.
———. 2006a. Historical Overview on Plant Neurobiology. *Plant Signalling and Behaviour* 1: 6–8.
Stanner, W. E. H. 1979. The Dreaming. In W. E. H. Stanner, *White Man Got Not Dreaming: Essays 1938–1973*. Canberra, Australia: ANU Press.
Steinfeld, H. et al. *Livestock's Long Shadow. Environmental Issues and Options*. Rome, Italy: FAO.
Stenhouse, D. 1974. *The Evolution of Intelligence: A General Theory and Some of its Implications*. London: George Allen and Unwin.
Stevenson, P. nd. The Seasons and Seasonal Markers of the Tiwi people of North Australia. Unpublished Manuscript.
Storch, I. 1993. Habitat Selection by Capercaillie in Summer and Autumn, is Bilberry Important? *Oecologia* 93: 257–65.
Struik, P. C., X. Yin, and H. Meinke. 2008. Plant neurobiology and green plant intelligence, science, metaphors and nonsense. *Journal of the Science of Food and Agriculture* 88: 363–70.
Sultan, S. E. 1996. Phenotypic Plasticity for Offspring Traits in *Polygonum persicaria*. *Ecology* 77: 1791–1807.
———. 2000. Phenotypic Plasticity for Plant Development, Function and Life History. *Trends in Plant Sciences* 5: 537–41.
Swearer, D. 2006. An Assessment of Buddhist Eco-Philosophy. *Harvard Theological Review* 99: 123–37.
Szerszynski, B. 2005. *Nature, Technology and the Sacred*. Oxford: Blackwell.
Tacitus. 1999. *Germania*, edited by J. B. Rives. Oxford: Clarendon Press.
———. 2004. *The Annals*, edited by A. J. Woodman. Indianapolis, IN: Hackett Publishing.
Tatia, N., ed. 1994. *Tattvartha Sutra*. San Francisco, CA: Harper Collins.
Tawhai, T. P. 2002. Maori Religion. In *Readings in Indigenous Religions*, edited by G. Harvey, 237–49. London: Cassell Academic.

Thistleton-Dyer, T. F. 1994. *The Folk-Lore of Plants*. Felinfach, UK: Llanerch Publishers.
Theophrastus. 1916. *Enquiry into Plants*, edited by A. Hort. In Greek with translation. Two volumes. Cambridge, MA: Harvard University Press.
———.1976–1990. *De Causis Plantarum*, translated by B. Einarson and G. K. Link. In Greek with translation. Three Volumes. Cambridge, MA: Harvard University Press.
———. 1993. *Metaphysics*, translated by M. van Raalte. Leiden, The Netherlands: E. J. Brill.
Thomas, C. D. et al. 2004. Extinction Risk from Climate Change. *Nature* 427: 145–48.
Thomas, R. F. 1988. Tree Violation and Ambivalence in Virgil. *Transactions of the American Philological Association* 118: 261–73.
Thompson, E. 2007. *Mind in Life: Biology, Phenomenology and The Sciences of Mind*. Cambridge, MA: The Belknap Press of Harvard University Press.
Tilman, D. et al. 2001. Forecasting Agriculturally Driven Climate Change. *Science* 292: 281–84.
Trenckner, V. ed. 1888–1925. *Majjhima Nikāya*. Four Volumes. London: Pali Text Society.
Trewavas, A. J. 2002. Mindless Mastery. *Nature* 415: 841.
———. 2003. Aspects of Plant Intelligence. *Annals of Botany* 92: 1–20.
———. 2004. Aspects of Plant Intelligence: An Answer to Firn. *Annals of Botany* 93: 353–57.
———. 2005. Green Plants as Intelligent Organisms. *Trends in Plant Sciences* 10: 413–19.
Tscharntke, T. et al. 2001. Herbivory, Induced Resistance, and Interplant Signal Transfer in Alnus glutinosa. *Biochemical Systematics and Ecology* 29, 1025–47.
Tucker, M. E. and J. A. Grim. 1994. *Worldviews and Ecology*. Lewisburg, PA: Bucknell University Press.
Tucker, M. E. and D. R. Williams, eds. 1997. *Buddhism and Ecology: The Interconnection of Dharma and Deeds*. Cambridge, MA: Harvard University Press.
Tudge, C. 2006. *The Secret Life of Trees: How They Live and Why They Matter*. London: Penguin.
Tull, H. W. 2004. Karma. In *The Hindu World*, edited by S. Mittal and G. R. Thursby, 309–31. London: Routledge.
Tunbrige, D. 1988. *Flinders Ranges Dreaming*. Canberra, Australia: Aboriginal Studies Press, AIAS.
United Nations Environment Progamme. 2007. *Global Environment Outlook GEO4, Environment for Development*. Nairobi, Kenya: UNEP.

Upadhyaya K. 1964. Indian Botanical Folklore. *Asian Folklore Studies* 23: 15–34.
Virgil. 1987. *The Aeneid*, edited by R. D. Williams. London: Allen and Unwin.
———. 1996. *The Eclogues and The Georgics*, edited by R. D. Williams. London: Bristol Classic Press.
———. 1999. *Eclogues, Georgics, Aeneid* I–VI, edited by G. P. Goold and translated by H. Rushton-Fairclough. Cambridge, MA: Harvard University Press.
———. 2000. *Aeneid VII–XII. Appendix Vergiliana*, edited by J. Henderson and translated by H. Rushton Fairclough. Cambridge, MA: Harvard University Press.
Vlastos, G. 1968. Does Slavery Exist in Plato's Republic? *Classical Philology* 63: 291–95.
———. 1973. *Platonic Studies*. Princeton, NJ: Princeton University Press.
Wandersee, J. H. and E. E. Schussler. 1999. Preventing Plant Blindness. *The American Biology Teacher* 61: 84–86.
———. 2001. Toward a Theory of Plant Blindness. *Plant Science Bulletin* 47: 2–9.
Wandersee, J. H. and R. M. Clary. 2006. Advances in Research Towards a Theory of Plant Blindness. In *Proceedings of the 6th International Congress on Education in Botanic Gardens at Oxford University*. London: Botanic Gardens Conservation International.
Warren, K. 2000. *Ecofeminist Philosophy*. Lanham, MD: Rowman and Littlefield.
Webster, C. 1966. The Recognition of Plant Sensitivity by English Botanists. *Isis* 57: 5–23.
Welch, C. A. et al. 1997. Constraints on Frugivory by Bears. *Ecology* 78: 1105–1119.
Whippo, C. W. and R. P. Hangarter. 2006. Phototropism, Bending Towards Enlightenment. *The Plant Cell* 18: 1110–19.
White, L. 1967. The Historical Roots of Our Ecological Crisis. *Science* 155: 1203–1207.
Whitt, L. A. et al. 2003. Indigenous Perspectives. In *A Companion to Environmental Philosophy*, edited by D. Jamieson, 3–20. Oxford: Blackwell.
Wiley, K. 2002. The Nature of Nature: Jain Perspectives on the Natural World. In *Jainism and Ecology: Nonviolence in the Web of Life*, edited by C. K. Chapple, 35–59. Cambridge, MA: Harvard University Press.
Williams, R. H. B. 1963. *Jaina Yoga: A Survey of the Mediaeval Sravakacaras*. New York, London: Oxford University Press.
Wilson, H. H., ed 1840. *The Vishnu Purana*. London: Trubner and Co.
Winter, N. 2000. *Tobacco Use by Native North Americans*. Norman, OK: University of Oklahoma Press.

Wolch, J. R. and J.-Emel. 1998. *Animal Geographies: Place, Politics, and Identity in the Nature-Culture Borderlands.* London: Verso.
WRAP 2007. *Understanding Food Waste Report Summary.* Online at www.wrap.org.uk/downloads/FoodWasteResearchSummaryFINALADP29_3__07.c0d5c6dc.pdf.
Zohary, M. 1982. *Plants of the Bible.* Cambridge, UK: Cambridge University Press.

INDEX

Aboriginal Australians, 101–3, 106–9, 112–13, 115, 168
Acacia tortilis, 112
Acaranga Sutra, 81, 83–85
Achuar, 115
action potential, 140–42, 147–48
Adnyamathanha, 101–2, 107
Adonis, 126
affinity, 81–82, 86, 90
agency, 1, 25, 62, 64, 105–6, 125, 128
agriculture, 31–35, 38, 64, 126, 131, 164–67
ahimsa. *See* nonviolence
Aislinge Oenguso, 124
alder, 124, 154
allegory, 63
allelopathy, 152–53
alloforms, 122
Ambrosia dumosa, 152–53
Amos, 62
ancestry. *See* kinship
Anglo Saxon Rune Poems, 128
animal-rights theory, 2, 157, 161
animism, 10, 100, 104–5, 114, 124–25, 128, 132, 134, 167–68
anthropocentrism, 5, 14, 20–27, 35, 48–50, 55–56, 62, 82, 119, 132, 167, 169
anthropomorphism, 105, 107

Apollo, 121
apple, 62
Aquinas, T., 41–42, 69–70
Aristotle, 7–9, 22–28, 41–48, 50–52, 57, 60, 67, 69–70, 140–41
ark. *See* Noah's ark
ash, 122, 124, 127–28
aspen, 131
atomism, 8, 46–50, 157
Augustine, St., 67–70
automatons, 46, 48, 52, 138, 144–45
autonomy, 2, 8, 11–14, 19, 23, 25, 30–32, 34–35, 37, 40–43, 48, 50–51, 61–66, 76–77, 80, 82–85, 90–91, 101–2, 105, 107, 111–14, 120, 128, 130, 132, 134–35, 156, 160–62, 169
auxin, 142, 147–48, 150

backgrounding, 1, 7, 10, 14, 19–27, 30, 47–49, 52–53, 57–58, 60–62, 67–68, 70, 83, 89, 93, 115, 120, 124, 135, 137, 157, 169
Bacon, F., 38, 43–46, 49–51
Bakhtin, M., 5, 9, 15, 162
Baluška, F., 12, 137, 147
bamboo, 74, 88, 97
barley, 65, 78, 152
Bashō, 97, 155, 162
Bauman, Z., 14

beech, 134
Bella Coola, 108
Bhagavadgītā, 73, 76–78, 80
Bhavachakra. See wheel of life
Bible, 25, 50, 55–70, 81, 119, 121, 131
bilberry, 161–62, 164
biofuels, 165–66
birch, 130
Bird-David, N., 10, 100, 105
Black Elk, 103
Blake, W., 157
Blodeuedd, 122
blood, 30, 39, 60–61, 70, 75, 102, 104, 112, 122, 125–27, 134, 160
Borassus flabellifer, 88
Braam, J., 142
Brahman, 75–6
brains, 6, 12, 26, 83, 139–40, 142, 147–48
breath, 8, 53, 57, 59–61, 68–69, 75, 82, 91
Buchloe dactyloides, 150
Buddha nature, 94–98
Buddhism, 9–11, 74, 86–98

Caledonian forest, 167
Callimachus, 126
Capra, F., 149
care, 11, 13–15, 26, 28, 30–31, 35, 56, 66, 76, 84, 100, 104, 115–17, 120, 123, 132, 134–35, 156, 159, 161, 167–69
castor oil, 66
cedar, 63, 65, 114–15
Cesalpino, A., 42–43, 47
Chan-Jan, 94–95
cherry, 123
children, 20, 44, 103, 106, 115, 125, 128
Chi-t'sang, 94
choice: between different actions, 4, 158, 160; in plants, 8, 12, 24, 33, 63, 139, 145–47, 149, 169
Chujin, 96–7
climate change, 159, 164–66
communication: between plants and humans, 12, 108, 110, 114, 125–26, 129, 133; between plants and other organisms, 12, 109, 147–48, 151–155, 169; capacity for, 10, 19, 24, 63, 108; within plants, 12, 139–42, 145–50
compassion, 66, 79–80, 92–93, 135, 160
computation, 144–46, 148, 154
Comte de Buffon, 37, 53
conflict (reducing), 11, 81, 85, 112, 114, 134, 160–61, 163–166
connection: ethic of, 11, 35, 77, 81, 86, 94, 100–2, 111, 114, 117, 120, 124, 156; familial, 123–24; ontological, 19, 29–30, 74–78, 80–81, 91, 102, 120, 124, 156; reconnection, 117, 132, 162; severance, 14, 68, 74, 91, 103, 163
consanguinity, 115
consciousness, 48, 50, 77, 79, 87, 91–92, 94, 106, 132, 135, 145, 149
consubstantiality, 11, 75, 94, 100, 104, 116, 121–24, 127, 132, 160
coolabah, 107
cor plantarum, 42
country, 101–2, 104, 106–110, 112–13, 116–17
crying. *See* tears
cultivation. *See* agriculture
Cuscuta, 146
cycads, 102, 108, 164
Cyperus rotundus, 152

Daiyi, N., 116
Damh, 132
Danyari, H., 101
Daphne, 122, 125
Darwin, C., 12, 137–47
Datura, 75
death: of nature, 159; of Phaëthon, 122; of plants, 1, 8–9, 23, 30, 56, 61, 66, 70, 78–79, 81–82, 87, 91, 110, 114, 122, 127–28, 168; of the first being, 121; of the young, 126
Della-Porta, G., 43–44
Demeter, 126–27
Descartes, R., 47–52
Descola, P., 105
Detwiler, F., 103, 106, 110

INDEX

Deuteronomy, 65
Deveraux, K., 116
dialogue, 9, 13–15, 80, 107–08, 129–33, 161–63, 166–69
Diodorus Siculus, 125
Dioscorides, 38, 41
Distant Time, 104
Dodona, 121
domination, 8, 20–26, 43–45, 49–50, 53, 56–57, 137, 157, 159, 167–69
dominion, 55–56, 61
Dreamings, 101–2, 104, 107, 117
Drosera rotundifolia, 140
Druids, 120, 132–35
dryads, 122–23, 125
Dutrochet, H., 138

Earth Bible, 56
eco-Paganism, 132–35
Eden, 60, 64
elm, 122–23
emergence, 149
Empedocles, 17–19, 21, 24, 34
enlightenment, 94–97
Erysichton, 126–27
ethics: environmental, 73, 86, 99, 119; human-plant, 80, 100, 156, 158; Jaina, 83; realization of, 85; relational, 2–4, 110, 168; virtue, 92–93
exclusion: ethic of, 11, 35; instrumental, 13; of nature, 19; of plants from life, 44; of plants from moral consideration, 14, 20–22, 26, 37–38, 46, 53, 86–87, 92, 156, 160; of plants from sentience, 47–49, 52–55, 74, 77, 87–95; philosophies of, 4–9, 17, 48–52, 71, 88, 98, 137, 157 (*see also* Aristotle, Bible, Buddhism, Descartes, Locke, Plato)

fig, 65–66, 123
fir, 119, 130–31
Firn, R., 145
first being, 121–23
flax, 103, 114
flesh, 30, 39, 55, 59–61, 75, 113, 117, 121–22, 128

flood, 59
flourishing, 8, 10, 13, 22, 31–34, 63, 79, 81–84, 108, 111–12, 129, 151, 153, 157, 162–63
foraging, 145–46
Forms, Platonic, 20
Frank, A., 138
free-will, 145, 149

Gagudju, 102, 108
Genesis, 52, 55–62, 64, 68
God, 8, 17, 20, 40, 42, 48, 50–51, 55–70, 75–76
Grimes, R., 135, 167
Grimnismal, 121–23, 127
Gruntman and Novoplansky, 150
Gunwinggu, 107

habitat loss, 159, 164–66
Hallé, F., 5–6, 26, 35
Hallowell, I., 10, 100, 105, 109
harm. *See* violence
Harvey, G., 10, 100, 105, 127, 132, 135, 168
hazel, 124
Heliades, 18, 125
heliotropism, 44, 46, 51
herbivory, reducing, 153–55
Hercules, 121
Hesiod, 18, 123
Hesychius, 124
heterarchy, 9, 11, 22, 57, 78, 105, 107, 109, 116, 121, 124, 132, 148–49
hierarchy: Aristotelian, 7–8, 22, 24–25, 35, 43–46, 49–50, 52; Buddhist, 10–11; Christian, 8, 52, 61–62, 64–65, 67–69, 121, 124; hierarchical value-ordering, 4–12, 21–22, 44, 53, 68, 157; Platonic, 19–22, 44, 49–50, 52
Hinduism, 73–81, 156
holly, 124
Homer, 125
Homeric Hymns, 123
Hönir, 122
human-nature dualism, 1, 7, 19–20, 47, 49, 61, 157, 167, 169

Hume, D., 158
Hyacinthus, 122
Hydrocotyle, 153
hyperseparation, 1, 99

Iga tree, 101–2, 107
immanence, 57
inclusion: of plants in totemic relationships, 116; of the natural world, 95; philosophies of, 9–13, 17, 35, 74, 101, 111, 115, 156 (*see also* Hinduism, Jainism, kinship, Theophrastus); within moral consideration. *See* moral consideration, sentience
instrumentalization, 8, 12, 14, 22, 25–27, 40, 43, 52–53, 61, 64, 66, 70, 85–86, 133, 156–57, 159, 164–65
intellect, 7–8, 23–24, 34, 67, 69–70, 77
intelligence, 12, 83, 135, 138–39, 141, 143–45, 147–49, 155–56, 158, 160–61, 169
intelligent plastic nature, 51
intentionality, 7, 32, 106, 144, 155
interdependence, 86–87, 103, 113, 116
intrinsic value, 56, 97, 158
Isaac, A., 116
Isaiah, 62
is-ought gap, 158

Jainism, 9–10, 73–74, 78, 80–91, 100, 111, 156, 160, 163
jangarla tree, 109
Jeremiah, 64
Jesus, 56, 65–66
jiva, 76–79, 81, 83
Job, 62–63
Jonah, 66
Jordan, W., 167–68
Jung, J., 47, 50
Jupiter, 121

Kalevala, 119, 129–31
karma, 78, 82, 84–85, 96, 98, 160
Katz, E., 56, 167–68
Kickapoo, 114

killing, 8, 65, 68, 70, 80, 84–85, 88, 91–92, 101, 111–13, 126–27, 130, 134, 160, 166
kinship, 11–12, 17–18, 24, 30, 58, 100–105, 110–11, 113–17, 120–27, 130–33, 137, 160, 167–69
kin selection, 155
Kohák, E., 1–4, 14, 157–58
Koyukon, 104, 109–10
Krishna, 74, 76
Kūkai, 95
kumara (sweet potato), 111, 115
kurrajong, 117
Kwaikiutl, 114

LaFleur, W., 94–97
Lakshmi, 74
Larrea tridentate, 152
laurel, 18, 121–22, 125
Law, 101–2, 107
learning, 145–46
lima bean, 154
Lincoln, B., 121–22
Linnaeus, C., 47, 52–53, 144
Locke, John, 49
Lódur, 122
lotus, 75, 77, 95
Lotus sutra, 94–96

Mabinogion, 122
Madhyamakahrdaya, 90
Mahābhārata, 73, 75–79, 83
Mahavira, 73, 81–83, 85
Mak Mak, 116, 168
Mancuso, S., 12
mangrove, 116–17
manners of speaking, 158
Manu Smrti, 79
Maori, 103–4, 111–12, 114–15
Mark, 66
Marvell, A., 113
Mathews, F., 2, 4, 48
Math Son of Mathonwy, 122
mechanism, 46–51, 138, 157
memory, 67, 83, 129

meristems, 147–49
metamorphosis, 12, 18, 53, 75, 100–107, 114, 121–22, 125
Mimosa: *pigra*, 168; *pudica*, 12, 52, 140–42
mind, 7, 12, 21–22, 24, 32–34, 37, 40–41, 48, 50, 60, 67, 69, 74–76, 83, 89, 91, 138, 148–49, 169
Minerva, 121
Minthe, 122
Mohawk, 113
Moltmann, J., 55–62
monologue, 5, 9, 15, 133, 161–62
moral consideration, 2–3, 13–15, 55–56, 71, 80, 84, 87–90, 92–94, 97–98, 108, 110, 116, 156–67
movement: absence of, 5, 41, 60, 69, 95, 138; growth, 138–40, 143, 145; in animals, 27; mechanical, 47, 52; nastic, 12, 46, 52, 138, 140–41; tropic. *See* heliotropism, tropism
mugwort, 128–29
mulberry, 123
Myrrha, 126
myrtle, 121, 127

Native North Americans, 102–6, 108–16
nature, concept of, 1–3, 19–20, 22, 40, 44, 48–50, 57, 132, 157, 159, 167
neem, 74
Neidjie, B., 99, 102, 104, 108–09, 112, 115
Nelson, L., 80
Nelson, R., 104, 109–10
nepes, 59–60
neurobiology, plant, 12, 147–48, 155
nerves, 52, 140–42, 147
Newton, I., 49
Ngarinman, 102
ngirrwat, 116
Ngulugwongga, 102
ngurlu, 117
Niebuhr, R., 56
Nikāyas, 87–88, 91
Nine Herbs Charm, 128–29

Noah's ark, 55, 58–59
nonviolence, 9–10, 79–81, 83–86, 89, 92–93, 111, 160
nutritive soul. *See* vegetative soul

oak, 63, 65, 119–24, 126–28, 130–31, 134
Odin, 122
Oglala, 102–3, 106, 110
Ojibwa, 105–6, 109, 116
olive, 44, 65, 121
O'odham, 64
Ophrys sphegodes, 153
Order of Bards, Ovates and Druids, 132–33
overconsumption, 163–66
Ovid, 121, 125

pain. *See* suffering
pandanus, 107, 117
Papatuanuku, 103
Parable of the Trees, 63
passivity. *See* exclusion from sentience, movement
Patimokkhaustta, 88
pea, 153
personhood, 10–14, 100–101, 104–107, 110–11, 114, 116, 124–35, 161–63, 167–68
persons. *See* personhood
phenotypic plasticity, 12, 143–46
philosophical ecology, 3, 101, 158
pipal, 75
Pitakas, 88
plantain, 129
plant blindness, 5
plantscapes, 3
plasmodesmata, 148
Plato, 7–9, 17–26, 29–30, 34, 41, 44–45, 48–52, 57, 60–61, 67, 70, 141
Pliny, 30, 37–43, 52, 120–21
Plumwood, V., 1–2, 6, 13–14, 19–20, 22, 49, 137, 159
Poetic Edda, 121–22, 127
pokeweed, 152

pomegranate, 65
poplar, 18, 121, 123, 125–26, 128
predation, 11–12, 100, 117, 146, 161
preference. *See* choice
presence, 2–3, 5, 9, 15, 74–75, 100, 107, 133, 161–62, 167
pre-Socratics, 17–19
Primavesi, A., 61–62
problem solving, 145–46
Psalms, 63
pseudocopulation, 153
Puketapu-Hetet, E., 103, 114
Puranas, 75, 77

radicle, 139, 141
Ranginui, 103
Ranunculus, 144
Ray, J., 38, 47, 50–52
reason: lack of, 21, 23–24, 42, 44, 52, 68–70; plant reasoning, 12, 81, 83, 145–46, 149, 155, 169; superiority of, 19–20, 24–25, 45
rebirth, 9, 76, 78–79, 82, 87, 91–92, 98, 100, 122
reincarnation. *See* rebirth
respect, 1, 13, 18–19, 35, 65, 71, 74, 79, 87, 99, 110, 112–14, 117, 119–21, 127, 131–32, 134, 156, 158–60, 163, 168
responsibility, 13–14, 26, 35, 55–56, 100, 112, 115–16, 120, 123–24, 134, 160
restoration, 14–15, 135, 166–69
Rhizobium, 151
rhizosphere, 151–52
rice, 75–76, 78, 152
ritual, 19, 64, 103, 110, 114, 135, 161–62, 167
Romulus & Remus, 123
root exudates, 151–52
Rose, D., 11, 101, 106–7, 111–12, 115, 117, 162, 169
ruach. *See* breath
Ryōgen, 96–97

sacred: groves, 119–21, 126; Hoop, 102, 106, 110; perfume, 77; plants, 74–75, 100, 114, 119–21, 127, 132; landscapes, 100; sacred/secular dualism, 61
Saichō, 95
samsara, 78–79, 87
scalae naturae, 24, 35
Schmithausen, L., 10, 74, 88–90, 97
Seeress's Prophecy, 122
self-recognition, 149–50, 153
sensory awareness, 12, 23, 26–27, 32–34, 46, 69, 76–77, 82–83, 87–89, 92, 108, 138–43, 148
sentience, 2, 7, 9–10, 12, 18, 35, 42, 44, 48, 52–53, 67–69, 74, 76, 79–84, 87–97, 102, 106–7, 110–11, 138, 156, 160–61(*see also* exclusion, sensory awareness)
Shinto, 95
Shitala, 74
signalling, 12, 138–50, 155
Singer, P., 2, 157
song: animist, 162; to cycads, 108; learning, 110; *Mower's*, 113; pagan, 125; of *Purusa*, 102; *of Solomon*, 62
soul: animal, 70; appetitive, 20, 23; Christian. *See nepes*; concept of, 110, 132; denial of, 8, 32, 68–69; as form, 22–23; Hindu/Jain. *See jiva*; human, 69–70, 110; intellectual. *See* rational; irrational, 67–68; nutritive. *See* vegetative; rational, 20, 24–25, 67–70; sensitive, 42, 70; tripartite, 20, 22–26, 41, 44, 47, 49, 51, 67–70; vegetative, 20, 23–24, 26, 32, 34, 41–44, 47, 49, 51, 70
sorghum, 109
spirits, 44–45, 105–6, 109–10, 124, 132–33
stewardship, 55–56
Stoics, 40
stringybark, 116
subjectivity: denial of, 24–25, 40, 53, 157; of animals, 2, 18; of country, 106–7; of plants, 13, 39, 41, 63, 116, 124–28, 130–32, 167, 169
suffering, 18, 21, 34–35, 39–40, 77, 79, 88, 91, 112, 125–28
superiority. *See* hierarchy

Sutta Nipata, 89
symbolism, 39, 62–63, 117, 121
synapse, plant, 147, 150
systems biology, 149

Tacitus, 120
Tanakh, 57
Tane-mahuta, 103
Tattvarta Sutra, 82–83
tears, 18, 126–27, 130, 157
thanchal tree, 109
Theophrastus, 7–8, 28–35, 38–41, 43–44, 47–48, 50, 156
Tiwi, 109
Tlingit, 104, 112, 114
tobacco, 114, 135
tomato, 146, 151–52
totem, 113, 115–16
transformation. *See* metamorphosis
tree-hugging, 134
Trees for Life, 167
Trewavas, A., 12, 143–46, 148, 156
Triphysaria versicolor, 153
tropism, 139
Tsimshian, 104, 112
tulasi, 74–75

Upaniṣads, 9, 73, 75, 78–79, 87

Veda, Rig, 75, 102, 121
vegetarianism, 80, 84
Venus, 121
Venus Fly Trap, 140–41
vine, 44, 65, 121, 123
violence, 9–12, 14, 34, 84–85, 91, 93, 111–13, 127, 130, 134–35, 160–64

Virgil, 123, 125, 127
voice, 1, 5, 9, 15, 24, 106, 108, 110–11, 130, 133, 161–62, 169
volatile organic chemicals (VOCs), 153–55
volition, 1, 7, 10–12, 20, 24, 48, 62, 76, 83–84, 95, 100, 105–7, 120, 124–25

wakan, 102, 106, 110
Warren, K., 7, 158
waste: avoiding, 85, 114, 146, 150; excretion, 27; of plant lives, 110, 160, 163–66
whakapapa, 103
wheat, 65
wheel of life, 87, 94
White, L., 55, 119, 124
Wik-Mungkan, 109
woolybutt, 116
Wulbulinimara, M., 117
Wuyaliya, 116

yam, 107, 112–13
Yanyuwa, 102, 108, 117
yew, 124, 128
Yggdrasil, 127–28
Ymir, 122–23
Yoga Sutra of Patanjali, 79
Yogavāsistha, 77

Zachariah, 63
Zeus, 121
Ziziphus spina-christi, 63
zoocentrism, 4–8, 10–12, 22, 26–27, 39, 42–45, 47–53, 57, 78, 83, 90, 119, 138, 141, 155, 157